Coastal Meteorology and Forecasting

Coastal meteorology encompasses a considerable range of small- and large-scale weather events that share underlying theoretical and practical principles. This book covers the foundational principles of coastal weather events and illustrates them through applications to real-world examples. A wide range of topics has been covered, from sea/land breeze circulations to low-level coastal jets and the interaction of cyclones with coastal features. This book represents an essential resource for upper-division undergraduates, graduate students, and researchers interested in coastal meteorology, oceanography, climatology, and atmospheric science. Readers will gain a solid conceptual understanding of meteorological phenomena that can be applied to coastal weather across the world and used to better predict coastal weather variations. This capacity to predict variations is necessary for mitigating climate change risk in coastal areas, which is an issue of current and pressing importance.

Dr. Wendell A. Nuss is Professor of Meteorology at the Naval Postgraduate School, California, where he has taught coastal meteorology for 40 years and led research on coastally trapped disturbances, coastal jets, and other coastal weather events. He has contributed extensively to coastal weather forecast training through the development of online training materials with the Cooperative Program for Operational Meteorological Education and Training (COMET) and workshops for US and other weather services.

Wendell A. Nuss has crafted a valuable contribution to the library of any active meteorologist. With his 40 years of experience researching and teaching coastal meteorology, Wendell has written a thoughtful and comprehensive book on the impact of coastal zones on our weather. This textbook covers the fundamental processes with careful explanations and clear and easy-to-comprehend graphics. The relevance of his work is magnified by the fact that the population of coastal zones has been increasing rapidly in recent decades, thus compounding the impact of coastal weather hazards.

Steven Businger, University of Hawaii

Coastal Meteorology and Forecasting by Wendell Nuss is an excellent review of coastal conditions, especially along the US West Coast. It has in one place all of the major developments from recent decades, with the references assembled by one of the major participants in this era. It is very nicely done with excellent figures. It was a pleasure to read this textbook.

Clive Dorman, Scripps Institution of Oceanography

Anyone living on or near a coast becomes aware that weather is “different” there. Experience teaches that daily patterns often repeat, but may be very localized, and that rare and locally concentrated extreme events can occur there as well. Nuss introduces the core conceptual and physical building blocks for understanding coastal meteorological physical processes. This includes clear mathematical representations and conceptual schematics. These graphs illustrate the key processes and structure, and their time evolution. They expose the core physical principles and how they are affected by a range of idealized coastal geographical conditions … Nuss’ deep personal experience in researching these physical processes and in teaching them extensively shines through. This book provides a solid foundation for understanding important coastal meteorological processes and their origins that equip the reader well both scientifically and practically. The coastal concentration of population and economic activities makes this book a useful background for any meteorologists responsible for research or weather prediction in such areas.It is also an excellent source for many others with a broad interest in the weather that affects their daily lives or businesses in coastal areas.

F. Martin Ralph, University of California, San Diego and Scripps Institution of Oceanography

Coastal Meteorology and Forecasting

WENDELL A. NUSS
Naval Postgraduate School

CAMBRIDGE
UNIVERSITY PRESS

CAMBRIDGE
UNIVERSITY PRESS

Shaftesbury Road, Cambridge CB2 8EA, United Kingdom

One Liberty Plaza, 20th Floor, New York, NY 10006, USA

477 Williamstown Road, Port Melbourne, VIC 3207, Australia

314–321, 3rd Floor, Plot 3, Splendor Forum, Jasola District Centre,
New Delhi – 110025, India

Cambridge University Press is part of Cambridge University Press & Assessment,
a department of the University of Cambridge.

We share the University's mission to contribute to society through the pursuit of
education, learning and research at the highest international levels of excellence.

www.cambridge.org
Information on this title: www.cambridge.org/9781009437264
DOI: 10.1017/9781009437271

When citing this work, please include a reference to the DOI 10.1017/9781009437271

First published 2026

A catalogue record for this publication is available from the British Library

Library of Congress Cataloging-in-Publication Data
Names: Nuss, Wendell Alan author
Title: Coastal meteorology and forecasting / Wendell A. Nuss.
Description: Cambridge ; New York : Cambridge University Press, 2026.
| Includes bibliographical references.
Identifiers: LCCN 2025036343 (print) | LCCN 2025036344 (ebook) |
ISBN 9781009437264 hardback | ISBN 9781009437271 ebook
Subjects: LCSH: Marine meteorology | Coasts
Classification: LCC QC994 .N87 2026 (print) | LCC QC994 (ebook)
LC record available at https://lccn.loc.gov/2025036343
LC ebook record available at https://lccn.loc.gov/2025036344

ISBN 978-1-009-43726-4 Hardback

Contents

Introduction 1

Coastal meteorology is the study of weather events in coastal regions that produce specific distributions and evolutions of winds, temperature, clouds, precipitation, visibility, and boundary layer structure that modify and define the coastal environment. These weather elements, while challenging in any region, become particularly difficult near coastlines, where processes associated with the coastal boundary add additional complexity. Although these weather events are usually not extreme, understanding the details of the structures that arise near coasts under even rather benign weather situations is of considerable interest to the coastal meteorologist. The seemingly small variations in the atmospheric structure near a coast can have profound impacts on numerous activities in the coastal region, such as aviation, recreational and commercial marine activities, and military operations. The focus of this book is to develop methods to understand, describe, and forecast the complex atmospheric structures that arise near coastlines.

1.1 Definition and Motivation

Any meteorological phenomenon occurring near a coastline that can be directly attributed to the presence of the coastal water/land boundary is considered to be a part of coastal meteorology. This broad definition encompasses a considerable range of weather events on both large and small scales. For example, mesoscale circulations, such as sea breezes, clearly owe their existence to the abrupt change in surface characteristics that occurs across a coastline. Likewise, larger-scale circulations such as tropical or extratropical cyclones can be substantially modified by the presence of a coastline to produce a variety of mesoscale structures. The mesoscale structures and mesoscale circulations that are forced by the coastal boundary are the primary focus of this book. In particular, understanding the dynamics of these mesoscale circulations as well as the larger-scale circulations that may trigger them or modify them in any significant way is the primary goal. The myriad of specific coastal circulations around the world will hopefully be united through a more general dynamic understanding of the basic forcing mechanisms and their interaction with larger-scale processes.

Operational numerical weather prediction models provide two important insights into our need to understand coastal phenomena. First, a look at the surface weather observations near almost any coastline reveals winds, clouds, and temperatures that do not seem to reflect the synoptic-scale conditions analyzed or predicted by larger-scale models. The ability to interpret these outlying observations as part of smaller-scale coastal circulation is of central importance to the coastal meteorologist, and being able to use that interpretation to add detail to a coastal forecast is critical to the marine forecaster, as large-scale models do not predict these structures. Second, advances in mesoscale modeling show that mesoscale models are highly capable of generating mesoscale structures within the coastal environment. However, the accuracy of mesoscale models in predicting the details of these circulations on any given day is not always known, and appropriate conceptual models of these coastal circulations are required to interpret the observations and to evaluate model forecasts within the coastal zone. The aim of this book is to develop appropriate conceptual models and to understand the basic physical mechanisms that force coastal circulations in order to adjust these conceptual models as a particular situation demands.

1.2 Scales and Types of Coastal Phenomena

As suggested in the definition of coastal meteorology, a broad range of spatial and temporal scales is important in coastal weather events. Typically, we are primarily interested in the mesoscale or smaller-scale aspects of coastal circulations, which arise from the smaller-scale structure of the coastline itself. Mesoscale coastal structures arise in two distinct ways. First, mesoscale dynamic forcing of the atmosphere by the coast can generate mesoscale circulations. In these situations, the synoptic-scale forcing is typically rather weak, and the coastal properties generate structural differences that dynamically force the atmosphere. For these types of circulations, key questions are what dynamics force the circulation and how does the synoptic scale forcing interact to allow or modify the coastal forcing. Second, mesoscale structures develop along coastlines through the interaction of the coast and a synoptic-scale circulation. These mesoscale structures are an artifact of the interaction for a given synoptic forcing and depend on the synoptic forcing in order to exist at all. For these types of circulations, the synoptic forcing tends to be strong, and the coastal properties alter the synoptic forcing to introduce or organize features within the synoptic-scale system. The coastal forcing is not fundamental to their existence but simply a modifying factor. Key questions are which coastal properties can influence larger-scale dynamics and how long or strong must this interaction be to cause a change in the synoptic system. An example of a mesoscale-forced circulation is the sea breeze, and an example of a synoptically forced circulation is the precipitation structures in a front as it passes a coast.

Figure 1.1 shows the mechanisms by which mesoscale coastal properties can produce mesoscale coastal weather events, given a spectrum of synoptic forcing. The top of the diagram shows the range of synoptic forcing from weak to strong, while the bottom of the diagram shows the types of coastal mesoscale circulations and their possible examples. The middle of the diagram represents the set of mesoscale coastal properties that influence the atmosphere in some manner. While weak synoptic forcing is often associated with dynamically forced coastal circulations, strong synoptic forcing can sometimes also support them. Similarly, strong synoptic forcing is often modified by coastal processes to produce mesoscale structures within the larger-scale system. However, this can also occur for weak synoptic-scale systems.

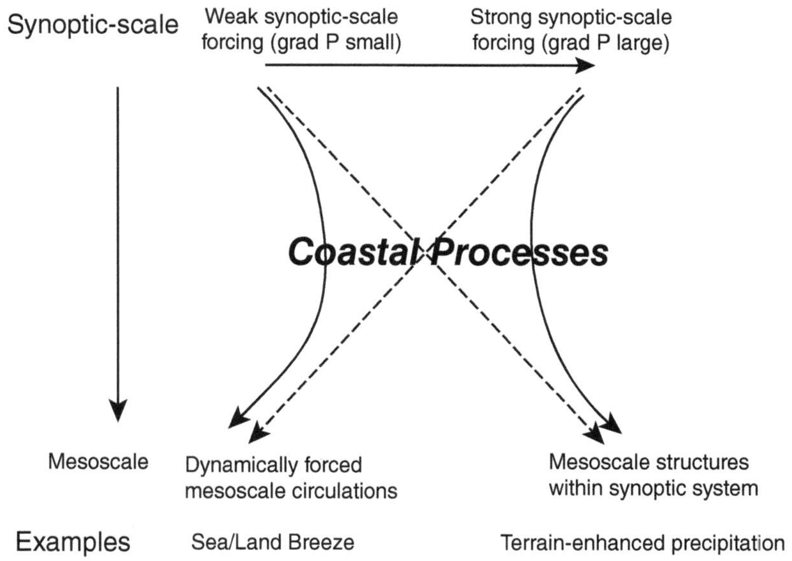

Fig. 1.1

Relationship between synoptic-scale forcing ranging from weak to strong shown across the top. Various coastal processes act on the synoptic-scale flow to generate mesoscale coastal circulations that are dynamically forced by the coastline or mesoscale structures within the synoptic-scale flow resulting from coastal effects. Examples on the bottom show circulations such as the sea/land breeze that can occur without synoptic-scale forcing to terrain-enhanced precipitation distributions, which would not occur without the synoptic-scale system.

1.3 Nature of the Coastal Boundary

To begin to understand the development and evolution of coastal circulations, it is first useful to consider the nature of the coastal boundary. The most simplistic representation of this boundary is a straight line with water on one side and land on the other. In this idealization, coastal mountains are not present, and there is little or no along-coast variation in properties along this coast line. With this simple coastline, surface characteristics vary across the coast, which produce frictional differences as well as surface heat and moisture flux differences. The differences in surface characteristics are sufficient to generate coastal circulations in the absence of other processes. Another important feature of some coastlines is the presence of coastal mountains, which produce a slope on the land side of the boundary. This change in surface slope from flat to inclined is also sufficient to generate coastal circulations that do not depend on the surface characteristic changes across the boundary. Of course, the real coastline includes considerable along-shore variability in surface characteristics and topographic slope. These along-shore variations contribute to local modifications of the circulations that arise under the more idealized coastlines as described earlier.

The impact of the coastal boundary occurs directly within the planetary boundary layer. The change in surface characteristics results in differing surface fluxes of heat, moisture, and momentum across the coastal boundary. The same synoptic-scale weather pattern can produce different boundary layer stratifications over land and water due to both differences in surface fluxes as well as differing flow characteristics over land compared to over water. The time evolution of boundary layer structure across a coastline is often critical to understanding the types of circulations that may arise. Strong stable stratification over land may strongly limit vertical mixing compared with that over water. This may account for differing inversion heights and their impact on the formation of low-level clouds. For many coastal circulation problems, the time evolution of boundary layer stratification and its impact on surface winds and vertical mixing within the boundary layer dictate the type of circulation likely to occur and its progression over time.

While the fundamental surface interaction doesn't change with location, the geographic specifics do impact the type and frequency of various coastal circulations. These differences occur due to climatological differences that set the background state within which coastal weather happens. East coasts of continents often have warm ocean currents, while west coasts of continents typically have much colder ocean currents. These ocean thermal differences influence both the cross-coast thermal structure and the overlying atmospheric conditions. The west coast is often characterized by a strong low-level temperature inversion, while the east coast tends to have weaker low-level stratification. These differences dictate the types of coastal weather events that are typically observed.

1.4 Idealized Coastal Circulations

The aforementioned simplification of coastal characteristics allows us to define essentially three idealized coastal meteorological problems (NRC, 1992). These idealized problems help to simplify our thinking about the more complex circulations that naturally arise along most coastlines. The three idealized problems can be classified as follows:

- thermally forced circulations;
- topographically forced circulations; and
- synoptic-scale forced coastal interactions.

These idealized problems describe the fundamental processes that are able to force coastal weather phenomena and are shown schematically in Fig. 1.2. Thermally forced circulations arise from surface heat flux differences across the coastline and are modulated by surface frictional differences. Pure thermally forced circulations occur with a flat or gently sloped coastal boundary and are often considered without along-shore variations. Examples of these types of circulations are the classical sea/land breeze and coastal fronts. Topographically forced circulations arise from a significant change in surface slope across the coast and do not depend on the surface characteristic changes that may also occur. However, the surface flux differences can modulate the pure topographically forced circulation. Examples of topographically forced circulations include lee vortices, flow-blocking effects, and cold-air damming. Synoptic-scale forced coastal interactions represent a class of smaller-scale phenomena that arise due to

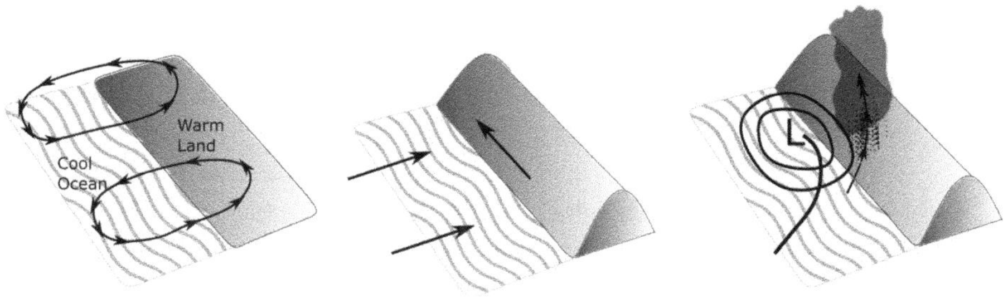

(a) (b) (c)

Fig. 1.2 Idealized types of coastal forcing. (a) Thermally forced circulations due to cross-coast temperature differences. (b) Topographically forced circulations due to flow interacting with coastal orography. (c) Synoptically forced circulations due to the interaction of larger-scale systems with the coast.

the passage of a larger-scale circulation across a coastline. These smaller-scale circulations occur when surface characteristics are able to modify larger-scale processes in a manner that spawns a circulation that may depend on both the coastline and the synoptic-scale system. Examples include the modification of fronts approaching or crossing a coastline to produce a coastal jet and tropical cyclones making landfall to spawn severe coastal thunderstorms. Under each type of idealized problem, a large list of locally observed phenomena can typically be found worldwide. It is beyond the scope of this book to examine every observed phenomenon that can be identified around the world. Instead, this idealization of the phenomena into three problems allows us to examine the fundamental dynamics governing each class of problems in order to understand its application to a particular observed event.

1.5 Mathematical Approach to Study Coastal Circulations

Since many of the coastal phenomena occur on smaller scales, they must be treated using mathematical models that include processes often eliminated when studying synoptic-scale systems. For example, quasi-geostrophic approaches will generally not be sufficient to describe or model coastal phenomena. Consequently, as with other areas of mesoscale meteorology, linear perturbation approaches to simplifying the equations and thinking about the phenomena are of great value. Complete treatments of mesoscale coastal circulations, of course, require sets of dynamic equations that contain all the relevant physical processes. However, we will concentrate on developing simple linear models that allow physical insight into the circulation as well as its interaction with larger-scale circulations.

To be successful in applying a linear perturbation approach to coastal circulations, the coastal circulation must satisfy several requirements. First, we must be able to assume that the time scale of the coastal circulation is short compared to the evolution of the background synoptic-scale circulation. This implies that the forcing must occur over a period of less than a day or so. Second, the spatial scale of the circulation must be small compared to the geostrophically dominated synoptic-scale. This implies that circulation-length scales must be less than a few hundred kilometers. Although these assumptions are required for the development of a linear perturbation model of a coastal circulation, it is possible to extend our simple model to include longer-lived and larger-scale effects where appropriate. However, the linear perturbation approach allows us to examine coastal circulations embedded in a slowly varying synoptic-scale background flow. Typically, the interaction in such an approach is down scale, meaning that the synoptic scale is allowed to modify the mesoscale but not vice versa.

In practical terms, this linear perturbation approach allows us to separate small-scale features on charts and consider the forcing of these features in isolation. An example of this approach is shown in Fig. 1.3, which shows both the synoptic-scale and small-scale perturbation pressure fields that might arise at a coastline. The large-scale pressure pattern (shown as solid lines) can be used to determine the geostrophic wind (solid arrows) and its evolution. The perturbation pressure pattern (shown as dotted lines) gives rise to small-scale pressure gradients to which the small-scale wind may or may not be given sufficient time to adjust geostrophically. If the wind responds geostrophically to the perturbation pressure, then the resultant observed perturbation wind might look like that shown by the dashed arrow that points along the coast in Fig. 1.3. If the wind does not respond geostrophically, the resultant observed wind might look more like that shown by the dashed arrow that points across the coast in Fig. 1.3. In general, we are interested in understanding the evolution and forcing of perturbation fields, such as pressure, wind, and temperature as well as their relationship to the corresponding large-scale fields. For example, we would like to know how strong the local acceleration might be due to a particular perturbation as well as what observed total wind might result. If we can model the behavior of the perturbations, then we can simply add this back into our known synoptic-scale background to find the observed evolution.

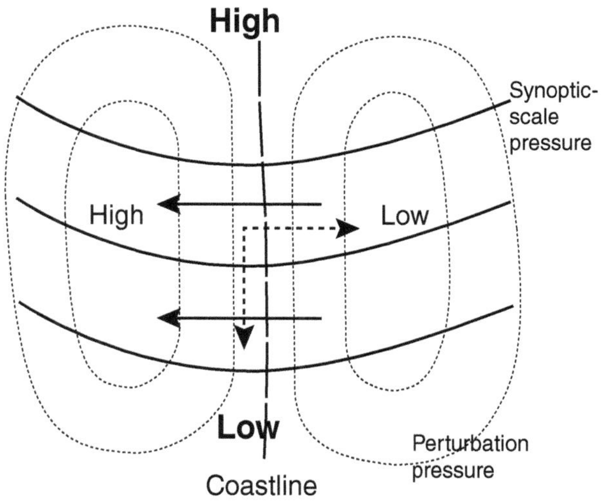

Fig. 1.3

Linear perturbation approach to separating the coastal circulation from the background large-scale circulation. The synoptic-scale background pressure is shown as thick solid lines and the associated winds as solid arrows. The perturbation pressure is shown as dotted lines and the associated winds as dotted arrows. The perturbation winds can be either ageostrophic toward low pressure or geostrophic, parallel to the perturbation pressure, as shown by the two dashed arrows.

In practice, we are often faced with knowing the total observed field and wanting to separate out the small-scale perturbation effects from a larger-scale field. This reverse problem is not easily done in practice in an objective manner. For example, scale separation techniques applied to objective analyses of observations may not always isolate the key perturbation features of interest. It is often useful to have a conceptual image of the idealized perturbation in mind when examining observations to see what smaller-scale forcing might be present. The aim in Chapters 5 through 11 is to develop this conceptual picture for various circulations.

The mathematical equations governing this perturbation approach can be derived from the basic governing equations of the atmosphere. We start from the primitive equations listed below:

$$\frac{du}{dt} = -\frac{1}{\rho}\frac{\partial p}{\partial x} + fv + \frac{uv}{a}tan\phi + F_x, \tag{1.5.1a}$$

$$\frac{dv}{dt} = -\frac{1}{\rho}\frac{\partial p}{\partial y} - fu + \frac{u^2}{a} + F_y, \tag{1.5.1b}$$

$$\frac{dw}{dt} = -\frac{1}{\rho}\frac{\partial p}{\partial z} - g + \frac{u^2 + v^2}{a}, \tag{1.5.1c}$$

$$\frac{d\theta}{dt} = \dot{Q}, \tag{1.5.1d}$$

$$\frac{1}{\rho}\frac{d\rho}{dt} = -\left(\frac{\partial u}{\partial x} + \frac{\partial v}{\partial y} + \frac{\partial w}{\partial z}\right), \tag{1.5.1e}$$

$$p = R\rho T. \tag{1.5.1f}$$

We can derive the linear perturbation equations by assuming that all the state variables can be decomposed into a larger-scale mean plus a small perturbation, as follows:

$$\begin{aligned} u &= u_o + u', \\ v &= v_o + v', \\ w &= w_o + w', \\ \theta &= \theta_o + \theta', \\ \rho &= \rho_o + \rho', \\ p &= p_o + p', \end{aligned} \tag{1.5.2}$$

where the subscript (o) denotes the larger-scale background flow and the prime represents the smaller-scale coastal circulation. Insert these quantities into the momentum equations, thermodynamic energy equation, equation of state, and continuity equation as follows:

$$\frac{d(u_o + u')}{dt} = -\frac{1}{(\rho_0 + \rho')}\frac{\partial(p_o + p')}{\partial x} + f(v_o + v') + \frac{(u_o + u')(v_o + v')}{a}tan\phi + F_x, \tag{1.5.3a}$$

$$\frac{d(v_o + v')}{dt} = -\frac{1}{(\rho_0 + \rho')}\frac{\partial(p_o + p')}{\partial y} - f(u_o + u') + \frac{(u_o + u')^2}{a} + F_y, \tag{1.5.3b}$$

$$\frac{d(w_o + w')}{dt} = -\frac{1}{(\rho_0 + \rho')}\frac{\partial(p_o + p')}{\partial z} - g + \frac{(u_o + u')^2 + (v_o + v')^2}{a}, \tag{1.5.3c}$$

$$\frac{d(\theta_o + \theta')}{dt} = \dot{Q}, \tag{1.5.3d}$$

$$\frac{1}{(\rho_o + \rho')}\frac{d(\rho_o + \rho')}{dt} = -\left(\frac{\partial(u_o + u')}{\partial x} + \frac{\partial(v_o + v')}{\partial y} + \frac{\partial(w_o + w')}{\partial z}\right), \tag{1.5.3e}$$

$$(p_o + p') = R(\rho_0 + \rho')(T_o + T'). \tag{1.5.3f}$$

If we ignore the spherical correction terms (those with the Earth's radius, a, in them) and assume that the background state is hydrostatically balanced:

$$\frac{dw_o}{dt} = -\frac{1}{\rho_o}\frac{\partial p_o}{\partial z} - g = 0 \tag{1.5.4}$$

and then expand the terms and drop any terms with the product of two perturbation (primed) quantities and the product of a perturbation with a derivative of the background ($u'\frac{\partial u_o}{\partial x}$), we get the following equations:

$$\frac{du_o}{dt} + \frac{du'}{dt} = -\frac{1}{\rho_0}\frac{\partial p_o}{\partial x} - \frac{1}{\rho_0}\frac{\partial p'}{\partial x} + fv_o + fv' + F_x, \tag{1.5.5a}$$

$$\frac{dv_o}{dt} + \frac{dv'}{dt} = -\frac{1}{\rho_0}\frac{\partial p_o}{\partial y} - \frac{1}{\rho_0}\frac{\partial p'}{\partial y} - fu_o - fu' + F_y, \tag{1.5.5b}$$

$$\frac{dw'}{dt} = -\frac{1}{\rho_0}\frac{\partial p'}{\partial z} - g\frac{\rho'}{\rho_o}, \tag{1.5.5c}$$

$$\frac{d\theta_o}{dt} + \frac{d\theta'}{dt} = \dot{Q} - w'\frac{\partial\theta_o}{\partial z}, \tag{1.5.5d}$$

$$\frac{1}{\rho_o}\frac{d\rho_o}{dt} + \frac{w'}{\rho_o}\frac{\partial\rho_o}{\partial z} + \frac{1}{\rho_o}\frac{d\rho'}{dt} = -\left(\frac{\partial u_o}{\partial x} + \frac{\partial v_o}{\partial y} + \frac{\partial w_o}{\partial z}\right) - \left(\frac{\partial u'}{\partial x} + \frac{\partial v'}{\partial y} + \frac{\partial w'}{\partial z}\right), \tag{1.5.5e}$$

$$p' = R(\rho' T_o + \rho_o T'). \tag{1.5.5f}$$

The equation of state (the aforementioned last equation) can be written as follows:

$$\frac{p'}{p_o} = \frac{\rho'}{\rho_o} + \frac{T'}{T_o}. \tag{1.5.6}$$

If we assume the background flow to be adiabatic and geostrophically balanced, then the horizontal equations of motion and continuity equations can be simplified to yield the following equations:

$$\frac{du'}{dt} = -\frac{1}{\rho_0}\frac{\partial p'}{\partial x} + fv' + F_x, \tag{1.5.7a}$$

$$\frac{dv'}{dt} = -\frac{1}{\rho_0}\frac{\partial p'}{\partial y} - fu' + F_y, \tag{1.5.7b}$$

$$\frac{dw'}{dt} = -\frac{1}{\rho_0}\frac{\partial p'}{\partial z} - g\frac{\rho'}{\rho_o}, \tag{1.5.7c}$$

$$\frac{d\theta'}{dt} = \dot{Q} - w'\frac{\partial\theta_o}{\partial z}, \tag{1.5.7d}$$

$$\frac{1}{\rho_o}\frac{d\rho'}{dt} + \frac{w'}{\rho_o}\frac{\partial \rho_o}{\partial z} = -\left(\frac{\partial u'}{\partial x} + \frac{\partial v'}{\partial y} + \frac{\partial w'}{\partial z}\right), \tag{1.5.7e}$$

$$\frac{p'}{p_o} = \frac{\rho'}{\rho_o} + \frac{C_v}{C_p}\frac{\theta'}{\theta_o}, \tag{1.5.7f}$$

where the temperature T has been converted into potential temperature θ in the equation of state. Further simplification can be made by using the Boussinesq approximation, where the density is assumed constant except in the buoyancy term, and the anelastic approximation, where the percentage range in potential temperature is small and the time scale is set by the Brunt–Väisälä frequency. This yields the following form:

$$\frac{du'}{dt} = -\frac{1}{\rho_0}\frac{\partial p'}{\partial x} + fv' + F_x, \tag{1.5.8a}$$

$$\frac{dv'}{dt} = -\frac{1}{\rho_0}\frac{\partial p'}{\partial y} - fu' + F_y, \tag{1.5.8b}$$

$$\frac{dw'}{dt} = -\frac{1}{\rho_0}\frac{\partial p'}{\partial z} + g\frac{\theta'}{\theta_o}, \tag{1.5.8c}$$

$$\frac{d\theta'}{dt} = \dot{Q} - w'\frac{\partial \theta_o}{\partial z}, \tag{1.5.8d}$$

$$\left(\frac{\partial u'}{\partial x} + \frac{\partial v'}{\partial y} + \frac{\partial w'}{\partial z}\right) = 0, \tag{1.5.8e}$$

$$\frac{p'}{p_o} = \frac{\rho'}{\rho_o} + \frac{C_v}{C_p}\frac{\theta'}{\theta_o}. \tag{1.5.8f}$$

These equations may be further simplified for specific applications where certain processes may dominate over others.

To assess the relative importance of key dynamical processes in a given flow situation, two important scaling parameters can be assessed. First, the relative importance of the Coriolis acceleration to other flow accelerations is assessed using the Rossby number, as follows

$$Ro\# = \frac{\frac{U^2}{L}}{f_o U} = \frac{U}{f_o L}, \tag{1.5.9}$$

where U is a characteristic velocity, L is a characteristic horizontal length scale, and f_o is the Coriolis parameter. This number is the ratio of the flow acceleration to the Coriolis acceleration and quantifies the tendency of the flow to be geostrophic. A small Rossby number implies that the Coriolis acceleration is large and the flow tends to be more geostrophic. When applied to coastal flows, this scaling parameter tells us whether the flow will adjust to the perturbation pressure gradient (geostrophic flow) or tend to flow down the pressure gradient in an ageostrophic manner. If the Rossby number is small, then the perturbation horizontal wind velocities can be approximated through geostrophic balance as given below.

$$-\frac{1}{\rho_0}\frac{\partial p'}{\partial x} = -fv'_g \tag{1.5.10a}$$

$$-\frac{1}{\rho_0}\frac{\partial p'}{\partial y} = fu'_g \tag{1.5.10b}$$

These relations show that the perturbation pressure gradient force, and Coriolis force balance to give geostrophic winds. Under this balance, the perturbation winds would (anti)cyclonically circulate around perturbation (high)low-pressure centers. This balance is often not achieved in many coastal circulations because the pressure perturbations often develop on relatively short time scales. By considering the thermodynamic structure in the vertical for geostrophically balanced flows, we can derive thermal wind equations that relate the geostrophic flow at one level to that at another level. The thermal wind equations for the perturbation flow are given by the following:

$$\frac{\partial u'_g}{\partial p} = \gamma\frac{\partial \theta'}{\partial y} \tag{1.5.11a}$$

$$\frac{\partial v'_g}{\partial p} = -\gamma\frac{\partial \theta'}{\partial x}, \tag{1.5.11b}$$

where $\gamma = \frac{R}{fP_0}\left(\frac{P_0}{P}\right)^{\left(1-\frac{R}{cp}\right)}$. These equations show that the geostrophic wind at one level is related to the geostrophic wind at another level based on the horizontal gradient of temperature (potential temperature) in the intervening layer. Again while this relationship may not apply to many coastal circulations, given that they are not in geostrophic balance, they do apply in some important coastal flows and are important to keep in mind when examining the vertical structure of coastal flows.

Another useful scaling parameter to characterize coastal flows is to compare the inertial forces to the gravitational forces acting on the flow. This is also equivalent to comparing the horizontal kinetic energy of the flow to the potential energy or work to displace the air vertically, which quantifies the importance of the gravitational force in modifying the flow. This is done using the Froude number, which can be defined for many applications in the form shown below.

$$Fr\# = \frac{U^2}{H^2N^2}, \tag{1.5.12}$$

where U is a characteristic horizontal velocity, H is a characteristic scale height, and N is the Brunt–Väisälä frequency. The numerator quantifies the kinetic energy of the horizontal flow, and the denominator gives the work required to displace the flow vertically through a height H given the gravitational potential energy of a stratified flow. A small Froude number implies that the work required to force the flow upward is large, which will tend to impede vertical motion and modify the horizontal flow. This assessment, using the Froude number, is important in considering the impact of topography on the flow and its ability to be forced over the topography.

The linear perturbation equations and simplification to them based on the Rossby or Froude # can be used to approximately model the behavior of many coastal circulations. These will be used in subsequent chapters to help elucidate key dynamical processes in various coastal circulations.

1.6 Exercises

1.1 Coastal circulations arise due to surface characteristic differences that occur across the land/water interface. List the surface characteristics that change across the coastline and how and to what depth they impact the atmosphere.

1.2 To appropriately apply the linear perturbation approach to describe coastal circulations, the coastal circulation must satisfy four characteristics relative to the synoptic-scale background. List and briefly describe these criteria.

1.3 Describe the direction that the perturbation flow will be oriented relative to a time-evolving pressure perturbation.

1.4 The perturbation pressure at the surface can be calculated from the vertical equation of motion (1.5.8c), assuming hydrostatic balance and integrating from the surface up to a level where the pressure perturbation vanishes.

- Write down the hydrostatic equation governing the perturbation pressure.
- Integrate the hydrostatic equation between the surface to a level z_i, where the perturbation pressure is assumed to vanish.
- Calculate the surface pressure perturbation given an average thermal perturbation of 2°C, $z_i = 1000\,\text{m}$, $\rho_0 = 1.3\,\text{kg/m}^3$, and a mean potential temperature of 288K.
- How much does the pressure perturbation change if the mean potential temperature was 305K, or the temperature rise was 1°C?

1.5 Calculate the vertical shear of the perturbation geostrophic flow in the surface to 850 hPa layer using the thermal wind equation given in eq. 1.5.11a. Assume that the perturbation potential temperature changes 2°C over a 100 km distance. Assume $f = 10^{-4}$ and $P_0 = 100000$ Pa. You can approximate γ for the layer using $P = 92500$ Pa and $R = 287$ and $c_p = 1004$. Then calculate the shear at a lower latitude where $f = 10^{-5}$. How much difference does it make?

Air–Land–Sea Interaction 2

Fundamental to understanding many coastal phenomena and their evolution is knowing how the atmosphere interacts with the underlying surface to produce horizontal and vertical structures. These differences in structure and surface interaction both force some circulations and modify others. In addition, the larger-scale background structure is established in part by the characteristics of, and interaction with, the underlying surface. This interaction with the underlying surface occurs in the lower portion of the atmosphere, which is referred to as the boundary layer.

2.1 Boundary-Layer Processes

The boundary layer is the lower portion of the atmosphere that is directly impacted by the underlying surface, both dynamically and thermodynamically. The boundary layer forms as air flows across the Earth's surface and experiences frictional drag produced by the fixed surface. In addition, heating of the Earth's surface produces vertical mixing through buoyancy to create a layer up to a height where the buoyancy becomes zero. While these processes, friction and thermal mixing, can produce layers of different depths in idealized flows, in the atmosphere there is usually a well-defined inversion or stable layer that caps the boundary layer at a height of z_i to limit the depth of both frictional drag and thermal mixing. The height of this capping inversion varies across both spatial and temporal scales and depends on several important factors.

First, the boundary layer is characterized by turbulent mixing processes that arise from both buoyancy and shear within the layer. The depth of the layer through which mixing and turbulence occur depends on the background stratification. The stratification is given by the vertical gradient of potential temperature $-\frac{\partial\theta}{\partial p}$, such that if this gradient is negative (positive), the layer is unstable (stable). While the atmosphere in general is stable to dry processes, the boundary layer can become unstable due to surface heat. If the layer is unstable, then the mixing of heat, moisture, and momentum occurs readily up through a fairly deep layer to the capping inversion. This consists of heat and moisture mixing upward through the layer and momentum downward through the layer to the surface where

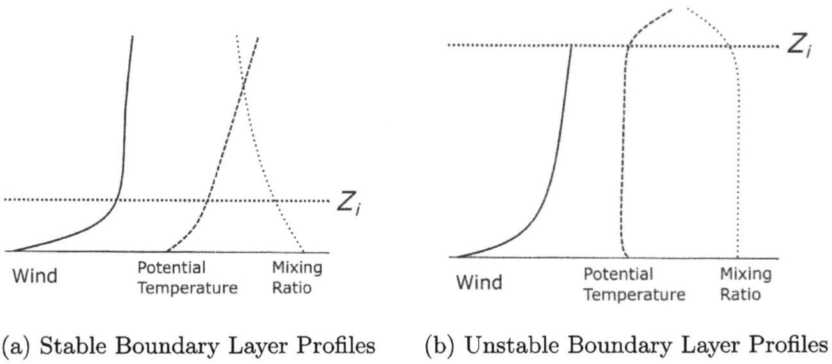

Fig. 2.1 Boundary layer profiles of wind (solid), potential temperature (dashed), and mixing ratio (dotted) for (a) stable and (b) unstable stratification. The top of the boundary layer is labeled as Z_i.

it is destroyed by friction. In this situation, the boundary layer can become deep. If the layer is stable, then turbulence is suppressed and mixing tends to be much more limited and is driven by shear turbulence more than buoyancy. In this case, the capping inversion tends to be rather low and the boundary layer is shallow. The typical boundary layer structure for stable and unstable stratification is depicted in Fig. 2.1. For stable stratification, the potential temperature increases upward, often most strongly near the surface. Wind speed may be relatively uniform above the surface but then decreases strongly near the ground where friction is strongest. The moisture profile can vary considerably, but often the moisture content or mixing ratio tends to maximize at the surface and decrease upward as the temperature decreases. For the unstable boundary layer, the potential temperature maximizes at the surface and decreases upward in the near surface layer. In the deeper boundary layer, vertical mixing tends to produce a relatively constant potential temperature in the vertical. The wind speed also tends to be relatively constant in the deeper boundary layer and then decreases more sharply in the near surface layer due to the surface friction. The moisture profile tends to be rather uniform through the mixed layer as a consequence of the buoyant mixing in the boundary layer. The humidity often decreases above the boundary layer.

The top of the boundary layer can be defined in various ways but essentially corresponds to the level at which turbulence drops to zero (or becomes very small). For the unstable, buoyant boundary layer, the top of the layer is often associated with an inversion or capping stable layer through which the mixing cannot penetrate. For the stable boundary layer, the top is generally defined as the level where friction vanishes or momentum mixing goes to zero. Given that the turbulence is mostly shear driven, the top of the boundary layer is often associated with the wind shear becoming small. These structures are illustrated in Fig. 2.1.

The depth of the boundary layer also depends on the large-scale flow characteristics. Within synoptic-scale low- and high-pressure systems, the

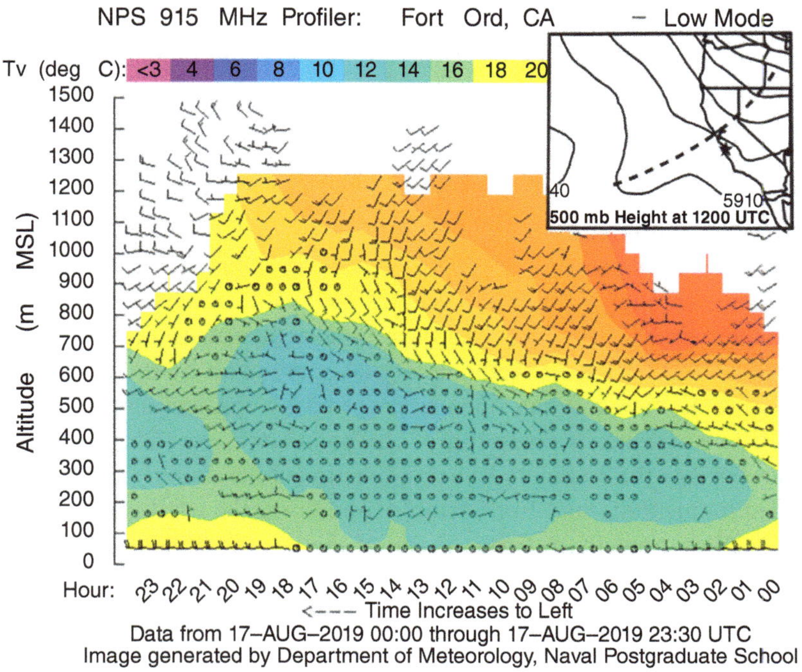

Fig. 2.2

Time series of wind profiler and virtual temperature from Monterey, CA. The marine boundary layer is the cool temperatures shown by the blue shading. The inset shows the 500 hPa geopotential height at 1200 UTC, August 17, 2019. The weak upper-level trough moves through the area, weakening the subsidence. As the subsidence weakens, the depth of the marine boundary layer rises from around 400 m to about 800 m over the day.

free atmosphere above the boundary layer can be characterized by large-scale ascent or subsidence. For example, in regions of high pressure, the atmosphere experiences subsidence, which tends to suppress the boundary layer depth by limiting the vertical mixing through turbulence. In regions of low pressure, ascent is forced by the synoptic scales, and the boundary layer depth can grow as the free atmosphere ascent aids the vertical mixing through turbulence. These impacts can be seen in a time series from a wind profiler along the coast near Monterey, California, as shown in Fig. 2.2. The lower layer, shaded blue in the figure, is cool ocean air, which is capped above by warmer temperatures shown in yellow and red colors. The warm air is the result of large-scale subsidence associated with the subtropical high pressure located west of the coast. On this day, the inversion starts at around 400 m (on the right side of the figure), and the cool layer deepens throughout the day to about 800 m. This rise in the inversion height is in response to the passing upper-level trough as depicted in the inset that shows the 500 hPa geopotential height. As the weak trough approaches, the subsidence weakens and the turbulent mixing in the layer penetrates higher without the opposing subsidence.

Critical to the forcing and evolution of coastal circulations is how the boundary layer processes and structure may evolve differently over land and water. The boundary layer structure evolves through interaction with the underlying surface as well as the character of the background flow. Surface interaction occurs through surface fluxes of heat, moisture, and momentum and is covered more completely in Section 2.2. The background flow is typically not uniform in the vertical, which impacts on the boundary layer processes. Given that friction occurs at the surface and decreases through the boundary layer, the wind profile is sheared with stronger winds at the top of the layer and weaker winds at the bottom. The winds at the top of the boundary layer are given by the background geostrophic flow and can often be assessed by considering the wind at 850 hPa as an approximate top of the boundary layer. The decrease in wind through the boundary layer is a function of the stability and the associated momentum flux. For unstable conditions, the wind shear tends to be small as momentum from the top of the layer is rapidly mixed down through the layer. For stable conditions, the wind speed can decrease very rapidly to near zero at the ground due to the lack of turbulent mixing in the layer. These wind shear profiles can impact on the stratification evolution if there is a background thermal gradient that leads to differential temperature advection through the layer. Warm advection can be stabilizing and cold advection destabilizing, when the advection is strongest at the top of the layer.

The boundary layer structure and evolution over land often vary diurnally as the ground warms and cools through radiative processes. This structure is illustrated in Fig. 2.3, which shows a near-surface stable boundary layer at night when the surface is radiatively cooled and a deepening mixed layer during the daytime hours as the ground is warmed. The layer between the capping inversion and the nighttime stable boundary layer is referred to as the residual layer, as this layer represents the structure produced within this layer the previous day. The depth of the mixed layer can be estimated by examining the early morning sounding and using the predicted daytime maximum temperature. The maximum temperature and surface pressure can be used to calculate the surface potential temperature, which can then be used to locate where on the sounding this potential temperature occurs in the vertical. For example, if the surface temperature is predicted to reach a temperature of 30°C, which corresponds to a potential temperature of 303K (assuming a surface pressure of 1000 hPa), then the depth of mixing will correspond to the level in the vertical where the potential temperature is 303K. This gives the depth of the mixed layer as illustrated in Fig. 2.4 for a morning sounding at Tampa, Florida. If the surface temperature rises to 30°C, the constant potential temperature of 303K is marked by the dashed red line in the figure. This suggests that the layer of mixing will reach a level of about 900 hPa. While the structure depicted in Fig. 2.3 is fairly typical, the vertical structure can be considerably more complex due to horizontal advection within the boundary layer.

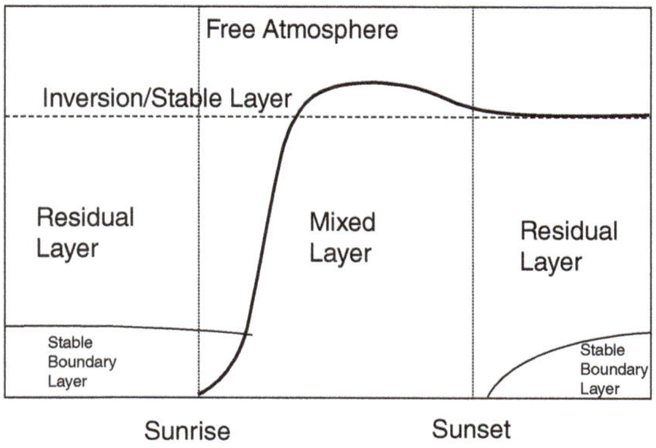

Fig. 2.3 Hypothetical boundary layer evolution over the land through a diurnal cycle. The thick dark line indicates the top of the mixing layer, which rises during the day as surface heating occurs. At night, after sunset, a surface-based stable boundary layer forms as a consequence of surface cooling at night.

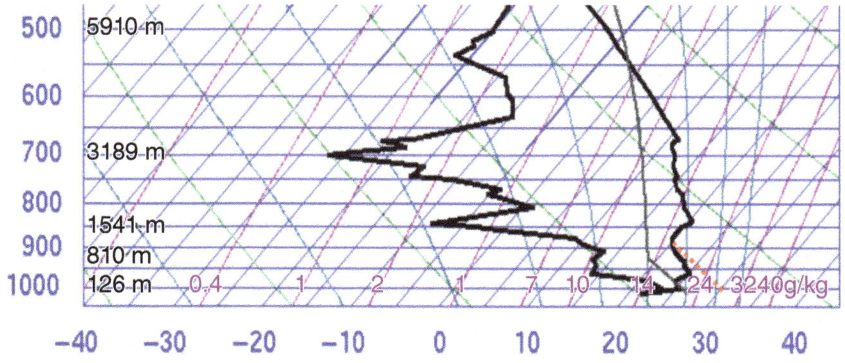

Fig. 2.4 The Skew-T diagram of the morning sounding (1200 UTC) from Tampa, Florida. If the surface temperature rises to 30°C at the time of maximum heating, the surface potential temperature becomes 303K, which mixes to a level of about 900 hPa, where 303K occurs on the sounding. The red dashed line represents the constant potential temperature of 303K on the plot.

The horizontal advection can produce internal boundary layers that may limit mixing to below the mixing depth suggested by the sounding.

The boundary layer structure and evolution over the ocean exhibit considerably less diurnal variation as the ocean surface temperature changes very little due to radiative processes. The structure is illustrated in Fig. 2.5, which shows the boundary layer as essentially well-mixed throughout the diurnal cycle. The capping inversion can rise slightly during the warmest part of the day as ocean surface temperatures rise due to the absorption of solar radiation. The amount of ocean surface temperature rise depends on several factors in the ocean. The depth of the thermocline, or depth of the underwater mixed layer, limits the increase in surface temperature as the

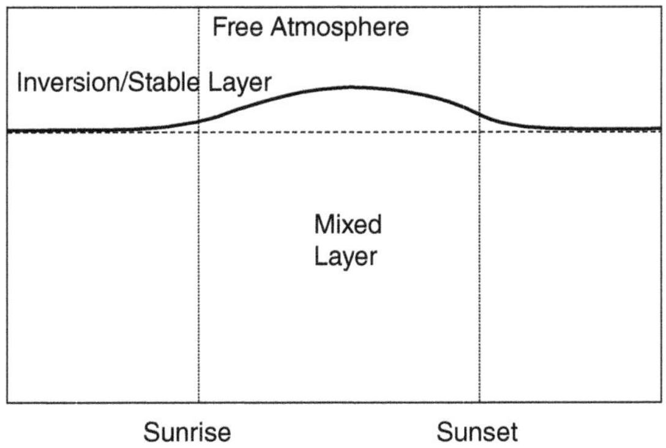

Fig. 2.5 Hypothetical boundary layer evolution over the ocean through a diurnal cycle. The thick dark line indicates the top of the mixing layer, which may rise during the day as some surface heating occurs. At night, after sunset, surface cooling doesn't occur and the layer stays well-mixed.

heat gets distributed throughout the ocean mixed layer. A shallow, near-surface thermocline is more likely to exhibit large diurnal temperature changes. The suspended particulates in the water also aid in the absorption of solar radiation. Clear water with few particulates will not absorb as much solar energy as more turbid water with lots of particulates.

2.2 Surface Interaction

Interaction of the atmosphere with the underlying land or ocean surface occurs through exchanges of heat, moisture, and kinetic energy or momentum. The exchange is given by a flux of heat, moisture, or momentum. These surface fluxes represent the rate of energy transfer and become primary drivers of the turbulent mixing processes that occur in the rest of the boundary layer.

The surface flux is determined by the difference in properties between the overlying atmosphere and the land or ocean surface. The temperature difference between the near-surface air and the ground or ocean surface determines the magnitude and sign of heat exchange between them. For example, if the ground is warm and the overlying atmosphere is cold, heat is transferred from the ground to the atmosphere through the surface heat flux. The moisture difference between the underlying surface and the atmosphere determines the magnitude and sign of the exchange of moisture between them. If the surface is moist (say over the ocean) and the atmosphere is dry, moisture is transferred from the ocean to the atmosphere through the surface moisture flux. The exchange of momentum is primarily downward into the surface as the atmosphere is generally moving relative to the ground or underlying ocean. This momentum transfer is the surface momentum flux and typically represents the drag that the surface imparts on the flow just above it. The magnitude and sign of the surface fluxes depend strongly on the characteristics of the underlying surface as well as the overlying atmosphere.

The surface fluxes are determined by the difference in properties at the surface and their profile just above the surface. While fundamentally a turbulent process, the fluxes can reasonably be represented through the bulk aerodynamic formula given as follows:

$$SH = c_p \rho C_h U_{10}(T_s - T_a) \quad (2.2.1a)$$

$$LH = L \rho C_m U_{10}(q_s - q_a) \quad (2.2.1b)$$

$$M = \rho C_d U_{10}^2, \quad (2.2.1c)$$

where SH is sensible heat flux, LH is latent heat flux, and M is the momentum flux. U_{10} is the 10 m wind speed, ρ is the air density, T_s is the surface temperature (ground or water), T_a is the surface air temperature, q_s is the surface moisture (specific humidity), q_a is the surface atmospheric moisture (specific humidity), and (C_h, C_m, and C_d) are exchange coefficients for heat, moisture, and momentum, respectively. c_p is the specific heat at constant pressure, and L is the latent heat of vaporization. These formula show that the magnitude depends on both the difference in temperature or moisture at the surface and the wind speed. For example, the heat flux can be large due to a large difference in temperature and relatively weak winds or a smaller temperature difference and strong winds. The exchange coefficients (C_h, C_m, and C_d) are represented as constants but do vary due to variations in the surface roughness and the near-surface stratification. Of particular interest is the impact of surface characteristics on the drag coefficient. Different land use characteristics lead to differences in the surface drag. For example, the drag over grass fields and forests is quite different. However, the difference between land and ocean is even more important. The surface drag over water is less than that of most land surfaces ($C_d = 1.5 \times 10^{-3}$ over water versus $C_d = 2 \times 10^{-2}$ over land for neutral stratification). The impact of these differences in the surface drag results in differing amounts of speed reduction over land and water. This reduction in wind speed causes the wind to turn toward low pressure by different amounts due to this friction.

The degree of wind turning toward low pressure can be estimated by considering the force balance between the pressure gradient force, the Coriolis force, and friction. Figure 2.6 depicts this three-way balance of forces. The angle of wind turning can be found if we consider that one component of the Coriolis force must balance a component of the frictional force. These can be written as follows:

$$C_r = fusin(\theta) \quad (2.2.2a)$$

$$F_r = C_d \frac{u^2}{h} cos(\theta), \quad (2.2.2b)$$

where these represent the components of the Coriolis (C_r) and frictional (F_r) forces orthogonal to the pressure gradient. The angle θ represents the degree of wind turning caused by the reduction in wind speed due to friction. The frictional force $\frac{\partial \overline{u'w'}}{\partial z}$ has been approximated by integrating through the boundary layer and using the bulk aerodynamic formula to get the vertical momentum flux. Equating the two forces and solving for θ, we get the following formula for the degree of turning caused by friction:

$$\theta = arctan\left(\frac{C_d u}{fh}\right). \tag{2.2.3}$$

This formula shows that as the surface drag C_d increases, the flow turns more toward low pressure. Higher winds turn more as well, and the effect is larger for shallow boundary layers. To illustrate this, we can calculate the degree of turning over water and land using drag coefficients of $C_d = 1.5 \times 10^{-3}$ and $C_d = 2 \times 10^{-2}$, respectively. For a boundary layer depth of 1000 m, a wind speed of 10 m/s, and a Coriolis parameter of 10^{-4}s^{-1}, we get 8.5° of turning over water and 63.4° of turning over land. These drag coefficients represent a very smooth surface over water and a very rough surface over land and therefore the degree of wind turning differences is amplified; however, the key point is that the turning toward low pressure over land is much more pronounced than over water.

Surface fluxes vary between land and the adjacent ocean even under the same background flow due to differences in the surface exchange coefficients as well as the change in surface properties. The surface fluxes depend on the surface temperature, moisture, and roughness compared to these quantities in the overlying atmosphere. For the heat flux, the ground temperature T_s varies diurnally and depends on the specifics of the ground itself. For example, dry sand compared to grass-covered moist soil has very different temperature increases for a given amount of solar radiation. The dry sand warms substantially more and so will produce a much larger surface heat flux. Over the ocean, the surface water temperature responds very little to solar radiation and its temperature remains fairly constant. The heat flux over the ocean is consequently rather constant through the diurnal cycle. Similar differences occur with the surface moisture flux. The moisture content in the ground can vary considerably, so the moisture flux may be downward with a moist atmosphere and dry ground or large upward for moist soil and a dry atmosphere. The moisture flux over the ocean changes little over time as there is an unlimited amount of water to be transferred into the atmosphere. The moisture flux over the ocean is primarily set by the moisture content in the atmosphere compared to the saturation mixing ratio associated with the ocean surface temperature. The momentum flux or drag is also quite different over land compared to over the ocean. The surface drag over the land is dictated by the land characteristics. Barren ground, low grass, forests, buildings, and so on determine the surface roughness and don't change (very often) for a given location. Almost all classes of land use produce greater surface

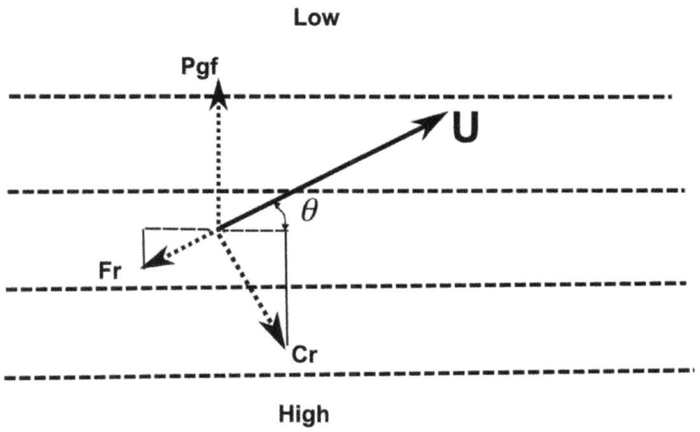

Fig. 2.6

The balance of forces that occurs between the pressure gradient (Pgf), Coriolis (Cr), and friction (Fr) causes the wind to turn toward low pressure. The angle of turning (θ) can be calculated by equating the components of the Coriolis force and friction force that are orthogonal to the pressure gradient (shown as dashed lines).

drag over land compared to water. This results in slower surface winds and more turning toward low pressure over land as shown in the calculation earlier. Over the water, the surface drag depends on the ocean wave characteristics. While overall this can be a complicated relationship, the first-order effect is that drag increases as wind speed increases due to the production of waves by the wind. Even under moderately strong winds, the surface drag over the ocean is typically less than most land surfaces, which gives stronger surface winds and less turning toward low pressure over the ocean.

2.3 Geographic Differences

The boundary layer and surface interaction vary widely due to large-scale weather patterns as well as specific details related to the surface in a given location. These weather pattern dependencies and how they evolve across the coast in a given flow regime are examined in considerably detail as we look at a specific coastal weather phenomenon. However, it is useful to note some geographic differences that occur in the boundary layer from a more climatological perspective. These more general tendencies to support certain types of boundary layer structures often dictate the types of coastal phenomena that can occur in a given location.

The east coasts of continents tend to be characterized by weaker stratification and warmer ocean temperatures. The large-scale ocean circulation produces poleward flowing currents along the east coasts of continents.

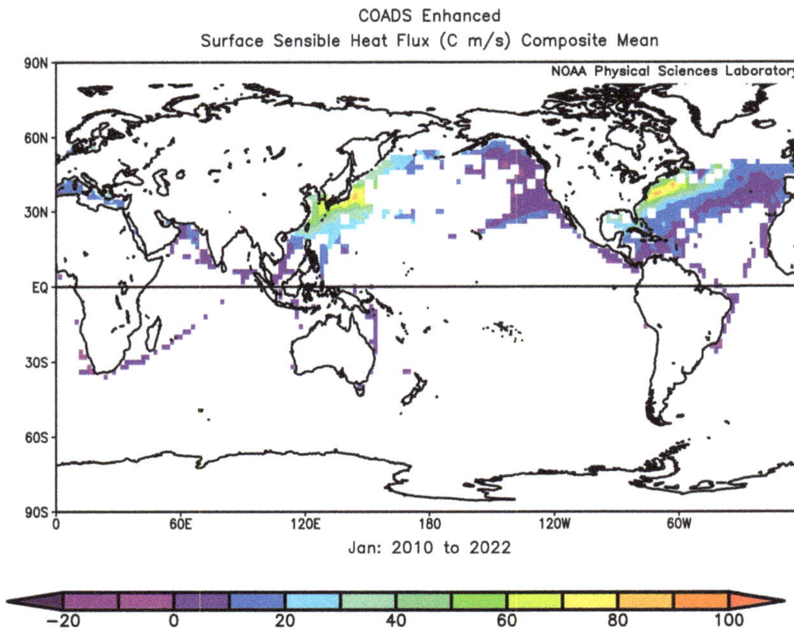

Fig. 2.7 Observed January climatology of surface sensible heat flux. Maximum surface heating occurs off the east coasts of North America and Asia. Weak or even downward fluxes occur off the west coasts of North America and Europe. Data/image provided by the NOAA Physical Sciences Laboratory, Boulder, Colorado, USA, from their website at https://psl.noaa.gov/.

This results in the transport of warm water poleward and the tendency for warm ocean temperatures even during winter. Given relatively warm surface temperatures, a positive heat flux is often observed as can be seen in the climatology of surface heat fluxes shown in Fig. 2.7. This has a tendency to produce a mixed layer with weak stratification. The climatological flow pattern in the atmosphere shows the tendency for long-wave troughs to occur along the east coasts of continents, which tends to produce weak mean ascent over the area that further promotes a weakly stratified boundary layer.

The west coast of continents has the opposite pattern to the east coast. Along the west coast, the large-scale ocean circulation produces a equatorward flowing current and colder ocean temperatures. This leads to downward heat fluxes and is shown in Fig. 2.7, which tends to stabilize the boundary layer. The climatological pattern in the atmosphere tends to produce long-wave ridging over the west coasts, which promotes subsidence over the region. Thus a well-defined inversion is often observed as the subsiding air warms and encounters the cold ocean below. Again these are general characteristics that can be disrupted by specific weather events, but they do provide some important constraints on the types of coastal events that may typically occur.

2.4 Exercises

2.1 On a skewT-logP chart, sketch the following situations:

- a well-mixed marine boundary layer;
- a stably stratified marine boundary layer; and
- a nighttime stably stratified land boundary layer.

2.2 Discuss the nature of the surface interaction (surface fluxes) and large-scale processes (divergence, advection, etc.) that contribute to and maintain the structures listed in the previous problem.

2.3 The boundary layer stratification can change due to surface heating/-cooling as well as thermal advection. Explain the expected change in stratification ($\frac{\partial \theta}{\partial z}$) for the following:

- a surface heat flux of 50 W/m^2;
- differential thermal advection with a uniform thermal gradient in the vertical and the wind speed increasing upward through the same layer.

2.4 Calculate the surface sensible and latent heat flux for air with a temperature of 10°C and a dewpoint of 5°C with a wind speed of 5 ms^{-1}. Assume the ocean temperature is 12°C. If the ocean temperature were 14°C, how much do the fluxes change?

2.5 As the wind speed increases, the surface roughness (drag) over the ocean increases due to the growth of ocean waves. If the drag coefficient increases by 50%, how much does the wind turning change compared to that given in the text?

3 Thermal Forcing of Coastal Circulations

The development of thermal gradients in the atmosphere is fundamental to the development of many circulations. The circulation around fronts, jets, and even larger-scale extratropical cyclones owes their existence to differences in thermal structure and the tendency of the atmosphere to respond to those differences. In considering the forcing of coastal circulations by the variations in coastal properties, thermal differences that arise across the coastal boundary are one of the primary physical processes. Temperature differences result in pressure gradients that can initiate or change the flow.

3.1 Development of Thermal Gradients

Temperature gradients can arise through a variety of processes, such as differing thermal advection by the wind, variations in vertical motion, and differential distributions of diabatic processes. Strong warm (cold) flow over the water and weaker warm (cold) flow over the land result in a cross-coast thermal gradient with warmer (colder) air over the water. Near the surface, this differential thermal advection can arise simply due to frictional differences between land and water, as noted in Chapter 2. Vertical motion results in adiabatic warming or cooling due to compression or expansion of the air. While this process can be very strong in the middle part of the atmosphere, low-level vertical motion does occur and can create thermal gradients. Perhaps most significant in coastal regions are thermal gradients that occur due to the diabatic processes associated with surface heat fluxes. Surface heating differences occur due to the differing surface energy balances that occur for water and land surfaces, which establishes a thermal gradient across a coastline.

To illustrate the process by which surface heating develops a thermal gradient in the coastal atmosphere, consider the energy balances that occur at the surface over water and over land. In both regions, the balance between solar radiation, infrared radiation, heat fluxes, moisture fluxes, and conduction into the subsurface layers acts to determine the surface temperature. While variations in the solar radiation reaching the ground occur due to clouds, aerosols, and other factors, these tend not to be

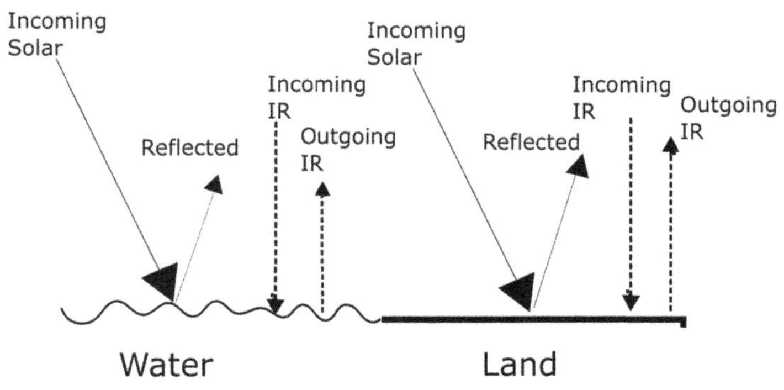

Fig. 3.1

Energy balance variations across the coast that lead to differing surface temperatures. Solid arrows represent the incoming and reflected shortwave or solar radiation. The dashed arrows are the longwave or infrared radiation that is coming from the atmosphere and being emitted from the surface.

significant when considering the difference in solar radiation reaching the water or land sides of a coastline. While the reflected solar radiation varies from over land versus over water, it is often larger over land. Assuming the land and water surface temperatures are identical, the difference between incoming and reflected solar radiation would suggest that the water should heat up more than land. However, the water has a very high heat capacity compared to the land, and consequently the temperature rise is much smaller over the water. This heat capacity difference is the primary contribution to the different surface temperature changes that occur across the coastline. The other factors can also vary across the coast, particularly the latent heat flux. Figure 3.1 shows the qualitative energy balances that occur over the water and adjacent land. The primary difference is that even though more solar radiation is reflected by the land, the much smaller heat capacity allows the absorption of solar radiation by the land to increase the surface temperature more than the water. This imbalance continues over the course of the day to lead to much warmer land temperatures and an associated cross-coast thermal gradient.

The differences in surface energy balances over a coastal region lead to both the tendency for a cross-coast thermal gradient as well as variations along the coast due to differing land and water characteristics. Figure 3.2 shows the horizontal distribution of surface temperature across a section of coastline over Central Florida from a model simulation. While the cross-coast thermal gradient is clearly dominant, variations in land characteristics lead to definite differences in the magnitude of this cross-coast gradient. Warmer surface temperatures occur to the south near Tampa, where the surface temperatures exceed 28°C as shown by the yellow shading. Further to the north along the west coast of Florida, the surface temperatures are cooler, around 26°C. This section of Florida is comprised

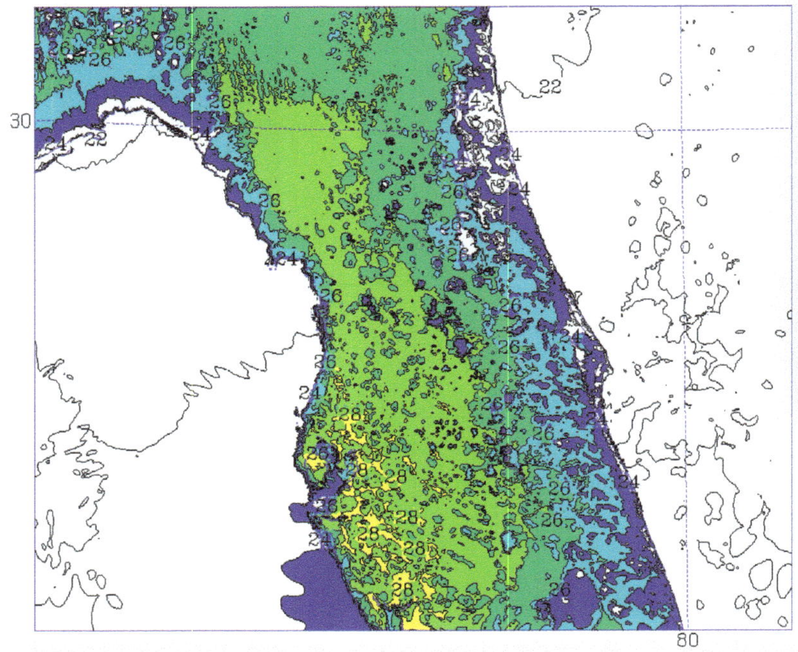

Fig. 3.2 Horizontal distribution of surface temperatures over a section of the Florida coastline from a high-resolution model simulation. The land characteristics vary over the region to give differing surface temperatures. Yellow and light green shading represent warmer temperatures than the blue shading. These heating variations along the coast result in differences in the cross-coast gradient.

of forests and other vegetated land cover, while further south near Tampa, there is considerably more urban or developed land covers. These differences will be explored more fully in Chapter 5 as they lead to complexity in the evolution of the coastal winds.

3.2 Dynamic Response to Thermal Gradients

In its purest sense, the coastal boundary is flat, and the large-scale flow is absent. The primary question to consider is how cross-coast thermal gradients in this situation result in the generation of a coastal circulation. Thermally driven circulations are very well understood from a theoretical basis and on larger scales essentially form the basis for adjustments in quasi-geostrophic flows. For example, the secondary circulation around a front, in a quasi-geostrophic sense, arises due to the changing thermal gradient forcing an ageostrophic flow. In the case of a front, the forcing comes from the large-scale flow field, altering the thermal gradient. In the case of a thermally forced coastal circulation, the thermal gradient is being forced through surface interaction processes.

To obtain a simple dynamic understanding of the development of a thermally forced circulation, consider the situation depicted in Fig. 3.3, and lets examine the circulation from the mathematical perspective of circulation. Mathematically, the circulation around a closed loop across the coast is defined as follows:

$$C = \oint \mathbf{V} \cdot \mathbf{dl} \approx VL, \tag{3.2.1}$$

where the vector velocity **V** is integrated around the loop **dl**. Approximately, the circulation is given by the product of the mean velocity V and the total path length of the circulation L. The circulation C defines the mean rotation in the two-dimensional plane containing the thermally forced circulation. The time rate of change in the circulation tells us how the forcing of the atmosphere over time alters the basic flow pattern. To obtain an expression for the time rate of change in the circulation C, we can start from the absolute momentum equation,

$$\frac{d\mathbf{V}_a}{dt} = -\frac{\nabla P}{\rho} - g \tag{3.2.2}$$

or

$$\frac{d\mathbf{V}_a}{dt} = -\frac{\nabla P}{\rho} - \nabla\phi, \tag{3.2.3}$$

where $\nabla\phi = g$ from the hydrostatic relation. We can then integrate the momentum equation around a closed loop that encompasses the entire coastal region where the temperature varies to obtain the time rate of change of the circulation.

$$\oint \frac{d\mathbf{V}_a}{dt} \cdot \mathbf{dl} = -\oint \frac{\nabla P}{\rho} \cdot \mathbf{dl} - \oint \nabla\phi \cdot \mathbf{dl}. \tag{3.2.4}$$

Note that the term on the left-hand side of the equation can be rewritten based on the product rule for differentiation as

$$\oint \frac{d\mathbf{V}_a}{dt} \cdot \mathbf{dl} = \oint \frac{d\mathbf{V}_a \cdot \mathbf{dl}}{dt} - \oint \mathbf{V}_a \cdot \frac{d\mathbf{dl}}{dt} \tag{3.2.5}$$

or

$$\oint \frac{d\mathbf{V}_a}{dt} \cdot \mathbf{dl} = \oint \frac{d\mathbf{V}_a \cdot \mathbf{dl}}{dt} - \oint \mathbf{V}_a \cdot d\mathbf{V}_a, \tag{3.2.6}$$

which leads to the following equation upon interchanging the order of integration and differentiation where possible

$$\frac{d}{dt}\oint \mathbf{V}_a \cdot \mathbf{dl} - \frac{1}{2}\oint d(\mathbf{V}_a \cdot \mathbf{V}_a) = -\oint \frac{dP}{\rho} - \oint d\phi. \tag{3.2.7}$$

The two terms on the right-hand side can be written as differentials by considering the nature of the vector dot product between the gradient and

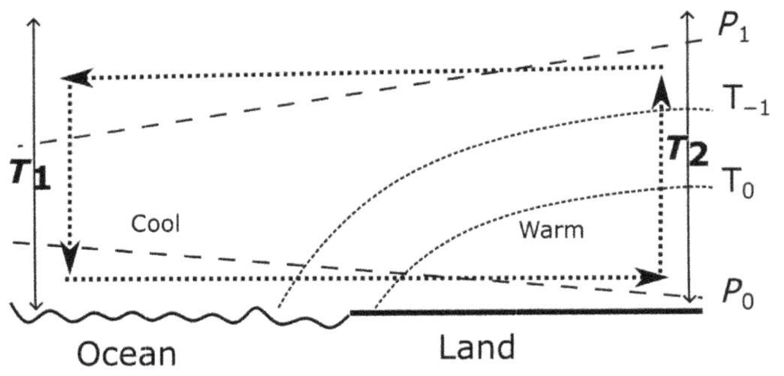

Fig. 3.3 Cross-coast thermal structure that generates a thermally forced coastal circulation depicted by the dashed arrows. Thin dotted lines are isotherms showing warming over the land and a cross-coast thermal gradient. The long dashed lines are two pressure surfaces, P_0 near the ground that slopes down over land and P_1 at the top of the warm layer that slopes down over the ocean. T_1 and T_2 represent the vertically averaged temperature over the ocean and land, respectively.

the line element. The last term vanishes because it is the integral of an exact differential around a closed path. The second term on the left side of the equation is also the closed loop integral of an exact differential and therefore vanishes. We are left with the following expression for the time rate of change of the circulation.

$$\frac{dC_a}{dt} = -\oint \frac{dP}{\rho}, \tag{3.2.8}$$

where the term on the right-hand side is referred to as the solenoidal term. The solenoidal term represents the generation of circulation (or vorticity) by the fact that the pressure and density surfaces are not coincident. If they are coincident, then this term vanishes as the right-hand side of the equation becomes an exact differential.

If we make use of the ideal gas law, then it is possible to write the forcing of the circulation in terms of temperature, which is the mechanism that gives rise to differing pressure and density surfaces. The circulation equation after substitution of the gas law ($P = \rho RT$) becomes as follows:

$$\frac{dC_a}{dt} = -\oint RTd\ln P. \tag{3.2.9}$$

This equation can be integrated along horizontal surfaces of constant pressure to yield

$$\frac{dC_a}{dt} = R\ln\left(\frac{P_o}{P_1}\right)(\overline{T}_2 - \overline{T}_1), \tag{3.2.10}$$

where the right-hand side is now in terms of the mean temperature difference across the coast between the P_o (lower) and P_1 (upper) pressure levels. Simply, the thickness between the pressure levels changes across the coast due to the temperature differences in the layers, which results in a height gradient or pressure gradient across the coast. This pressure or height gradient forces the air to accelerate toward lower height or pressure to force the circulation. This is a very classic result covered in numerous dynamics textbooks (e.g., Holton, 2004). The key point is that if a thermal gradient develops (baroclinic structure), the air must respond to the pressure gradient to force a circulation. If the thermal gradient stays fixed over time, the atmosphere will simply adjust to the pressure gradient due to the Coriolis force, and the winds will tend toward geostrophic balance. Hence, the initially cross-coast forcing would result in geostrophically balanced along-coast winds. The time scale of this adjustment is given by the inertial period, given that the thermal structure does not change due to the induced circulation.

This basic dynamic forcing is important for considering the thermal aspects of coastal circulations. Obviously, the biggest problem is not understanding the dynamic forcing but understanding the evolution of the thermal structure over time and its impact on and modification of the evolving coastal circulation. It is this aspect of the forcing that will be considered more carefully in Chapter 5.

3.3 Time- and Space-Scale Considerations

As noted above, the role of the circulation induced by thermal gradients is to create a flow that is in balance with the thermal gradient. This adjustment happens due to a combination of Coriolis acceleration of the cross-isotherm flow and adiabatic heating and cooling by the vertical motion. If the thermal gradient stays fixed, this adjustment process will eventually result in a flow that is in geostrophic balance with the pressure gradient produced by the thermal gradient. This adjustment process is ubiquitous to larger-scale flows in the mid to higher latitudes, where the Coriolis force is of significant magnitude and is referred to as geostrophic adjustment. At lower latitudes, where the Coriolis force is small, only the vertical parts of the circulation can change the thermal gradient to weaken the circulation. This process is considerably less efficient in bringing the flow into balance.

While the complete adjustment to geostrophy happens more quickly due to the vertical circulation altering the thermal gradient, the effect of the Coriolis force to rotate the wind can be understood in terms of the inertial cycle. Figure 3.4 illustrates the evolution of the wind over time due to the impact of the Coriolis force acting on the wind. The Coriolis force

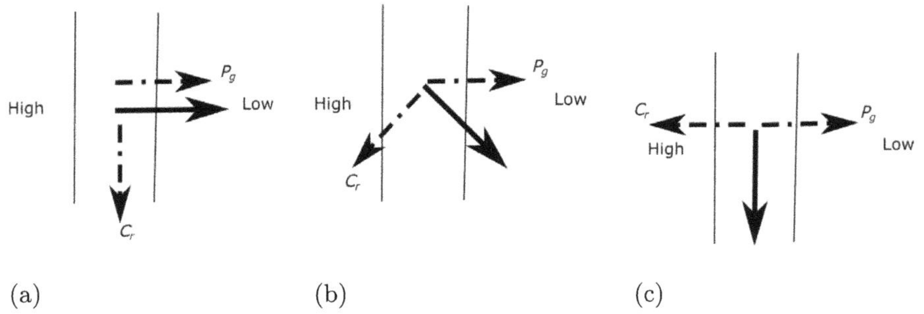

Fig. 3.4 Evolution of winds over the inertial cycle. The dot-dash arrows represent the pressure gradient (P_g) and Coriolis (C_r) forces, and the solid arrow represents the wind. Initially, the pressure gradient force and Coriolis force are perpendicular to each other with the wind going toward low pressure. As the Coriolis force acts on the wind, the wind turns to the right (middle panel) and eventually is parallel to the pressure contours (geostrophic balance). (a) Initial force balance, (b) Intermediate period force balance, and (c) Final force balance

acts to the right of the flow (in the Northern Hemisphere) and initially is orthogonal to the pressure gradient force that has produced a completely ageostrophic down-gradient flow. This force starts to rotate the wind to the right such that part of the Coriolis force balances the pressure gradient force and part acts to turn the flow more to right. This turning will continue until the wind is parallel to the pressure contours (perpendicular to the pressure gradient), at which point the Coriolis force and pressure gradient force balance. The time to complete this cycle is given by $t = \frac{2\pi}{f}$, which is the inertial period. At 30 N, the inertial period is approximately 24 hours. This represents the maximum amount of time required for the wind to adjust geostrophically. As noted above, the vertical branches of the circulation illustrated in Fig. 3.3 act to reduce the temperature gradient and associated pressure gradient. This reduction in the pressure gradient causes the adjustment to geostrophy to occur more quickly than the inertial period.

The consideration of geostrophic adjustment of the winds to a cross-coast thermal gradient (and pressure gradient) has some important implications for the time evolution of thermally driven circulations. Given the strong dependence of the thermal gradient on the solar radiation, coastal thermal gradients typically undergo growth and decay and then reversal over the diurnal cycle. Consequently, the thermal gradient exists for well less than the inertial period, and complete adjustment to geostrophy is rarely observed. However, the impact of the Coriolis force is clearly observed in thermally driven cross-coast flows as the wind tends to rotate to the right over the course of the diurnal cycle. Given the time-varying thermal gradient as well, the exact evolution of the wind is considerably more complex than geostrophic adjustment to a fixed thermal gradient. This time evolution and its impact on the cross-coast flow evolution and penetration will be considered more completely in Chapter 5.

3.4 Exercises

3.1 Consider the forcing of thermally driven circulations from the idealized perspective given in the text.

- Write down the expression governing the evolution of a thermally driven circulation.
- Calculate the change in circulation that would result if the mean temperature in the 1000 to 850 hPa layer across the coast were 2°C and 10°C higher over the land than the ocean.
- If this thermal change is assumed to occur gradually (linearly) over a 12-hour period, calculate how much the wind should increase over this period of time. Assume that the wind speed is related to the circulation through the length scale of the circulation ($C = Vl$). Assume that a reasonable length scale is 100 km.
- Comment on the reality of your results from the above calculation and describe what aspects of the real atmosphere are missing from this simple model.

3.2 Given the same radiative heating over land versus water, how much more will the surface temperature rise over land compared to water? Assume that the heat capacity of the water is 4000 J/kg/C and for the soil is 800 J/kg/C. Assume a radiative heat flux of 500 W/m2 and that the heating penetrates only to 1-m depth with no other processes impacting the heat balance. (Note that the heat balance in the soil and water is more complex.)

3.3 Calculate the inertial period at 30 and 60 N latitudes and compare this time period to the length of daylight at these locations for both winter and summer.

3.4 For the locations in the previous question, how far will the winds have turned by the end of the daylight hours?

4 Topographic Forcing of Coastal Circulations

Many coastlines are characterized by mountains or significant slopes, which can impact many aspects of the coastal circulations and associated weather. While thermal forcing can generate circulations in the absence of a background flow, topographic forcing requires the interaction of a background flow to produce circulations. Topographic impacts on the flow are not unique to the coastal region, and many of these effects occur with inland mountains as well as along the coast. We will focus on the fundamental aspects of topographic forcing in this chapter and look at specific impacts in coastal regions in Chapters 6 through 11.

4.1 Basic Flow Interaction with Topography

In considering the interaction of a flow with coastal topography, several general factors must be considered. First, what is the nature of the topography itself: an isolated mountain, a long, high mountain barrier, a gradually sloping coastal plain, or any other. Each type of coastal topography can dictate a particular response for the same synoptic-scale flow regime. Second, the nature of the incident flow plays a big role in determining the response to a given topographic feature. This dynamically controlled process will be examined first to begin to get a general sense of how an air flow interacts with mountains.

In considering the general nature of flow interacting with topography, the most basic question is whether the air will go up and over a mountain or be forced around it. Let's consider this problem using a simple analogy: rolling a marble over an incline. In this simple physics problem, we need to consider the height of the incline (mountain height) and the velocity of the marble approaching the incline (momentum). The marble will go up and over the incline if its initial momentum is sufficiently large to overcome the gravitational potential energy of the incline. In fact, the ratio of the kinetic energy to the gravitational potential energy gives the correct measure of whether the marble goes over the hill. If $\frac{KE}{PE}$ equals 1, then the marble just makes it to the top of the hill, if greater than 1, it goes over and if less it rolls back down. Although this is a highly simplified example, the same basic principle applies to atmospheric flows.

For atmospheric flows, the gravitational potential energy is represented through the static stability, and the momentum or kinetic energy is the incident wind velocity. For a statically stable atmosphere, the restoring force acting on a lifted parcel is the difference between the ambient lapse rate and the dry adiabatic lapse rate after lifting a parcel from its initial height to some new height such as the top of the mountain (h_m). This restoring force is given directly from the Brunt-Vaisala frequency ($N = (\frac{g}{\theta_o}\frac{\partial\theta}{\partial z})^{\frac{1}{2}}$). Consequently, the condition of whether a given air parcel will go up and over a mountain is given heuristically as ($Fr = \frac{U}{h_m N}$), which is a Froude number associated with the mountain. If this number is greater than 1, then the air parcel will make it over the mountain; if less than 1 then it will not, and if equal to 1, then it reaches the mountain top with zero velocity. For low Froude numbers (less than 1), this simple physical reasoning suggests that the flow is essentially blocked by the topography and must either go around or be turned back.

4.2 Flow Blocking and Flow around Mountains

Certain properties of flow blocking can be demonstrated by applying integral constraints to a hypothesized, simple, steady flow. Consider the stably stratified flow over a step illustrated in Fig. 4.1. Suppose the flow far upstream from the step is characterized by an incident flow speed ($-U_o$) and constant stratification (N). We can ask, under what conditions can we get dynamically self-consistent solutions for the flow very far downstream from the step? As suggested in the heuristic reasoning in Section 4.1, downstream solutions likely do not exist if the Froude number is less than one. A Froude number less than unity can be shown to be a sufficient condition for downstream solutions not to exist and the flow to be blocked. While this criterion may not strictly be a necessary condition for flow blocking, it agrees well with numerical experiments which suggest that flow blocking begins at Fr = 1.

To show that a Froude number of one is a sufficient condition, consider the inviscid, Boussinesq, adiabatic momentum equation.

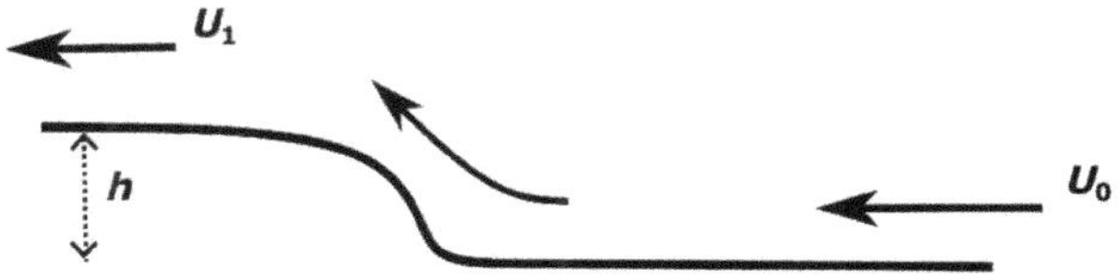

Fig. 4.1 Flow interacting with simple topographic step. Incident flow has a velocity u_0 and is forced up a step of height h. The downstream velocity is u_1.

$$\frac{d\mathbf{V}}{dt} = \frac{-1}{\rho_o}\nabla p' + g\frac{\theta'}{\theta_o}\mathbf{k}, \tag{4.2.1}$$

where p' and θ' are deviations from the upstream profile of p and θ. We can form a Bernoulli equation by considering the energetics of the flow and integrating the momentum equation along a surface streamline following a parcel.

$$\int_0^1 \frac{d\mathbf{V}}{dt} dS = \int_0^1 \frac{-1}{\rho_o}\nabla p' ds + \int_0^1 g\frac{\theta'}{\theta_o} dS, \tag{4.2.2}$$

where 0 is a point upstream and 1 is a point downstream. Note that $dS = \mathbf{V}dt$ and that $\nabla p' = \frac{\partial p'}{\partial x}$ due to the fact that the flow is hydrostatic. Additionally, θ' varies only in the vertical due to the flow being adiabatic and integrating along a streamline. Given these observations about the situation, our integral equation can be written as

$$\int_0^1 \frac{d\mathbf{V}}{dt}\mathbf{V}dt + \frac{1}{\rho_o}\int_0^1 \frac{\partial p'}{\partial x}dx - \frac{g}{\theta_o}\int_0^1 \theta' dz = 0 \tag{4.2.3}$$

which can easily be evaluated to be

$$\Delta\frac{1}{2}V^2 + \frac{1}{\rho_o}\Delta p' - \frac{g}{\theta_o}\int_0^1 \theta' dz = 0, \tag{4.2.4}$$

where the Δ represents the difference between the downstream and upstream points. The last term can be written as $\int N^2 \Delta z dz$ if constant stratification is assumed. This comes directly from the definition of the Brunt-Vaisala frequency $N^2 = \frac{-g}{\theta_o}\frac{\partial\theta'}{\partial z}$. Thus our Bernoulli equation is simply as follows:

$$\Delta\frac{1}{2}V^2 + \frac{1}{\rho_o}\Delta p' - \frac{1}{2}N^2(\Delta z)^2 = 0. \tag{4.2.5}$$

This expression can be evaluated for the streamline that hugs the ground at the upstream and downstream locations from the step to get the following:

$$\frac{1}{2}U_1^2 - \frac{1}{2}U_o^2 + \frac{p_1'}{\rho_o} - \frac{p_o'}{\rho_o} + \frac{1}{2}N^2h^2 = 0, \tag{4.2.6}$$

where h is the step height. Since no pressure perturbations occur far upstream from the step, the term p_0' is zero and the equation is

$$U_1^2 - U_o^2 + \frac{p_1'}{\rho_o} - N^2h^2 = 0. \tag{4.2.7}$$

If we can obtain an independent expression for p_1', then we have the solution for the wind downstream given the upstream conditions. If point 1 is far downstream, then we might expect p_1', to vanish. Considering this simple situation where $p_1' = 0$, then

$$U_1^2 = U_o^2 - N^2h^2. \tag{4.2.8}$$

This expression implies imaginary solutions for U_1 if N^2h^2 exceeds U_o^2, which presumably corresponds to the flow being blocked. Thus, the Froude number being less than unity $Fr = \frac{U_o}{Nh} < 1$ gives a first-order sufficient condition for flow blocking.

This simplified mathematical reasoning can be extended to incorporate the effects of when the downstream pressure is allowed to vary. Although we do not know the downstream pressure, we can examine this more general condition and derive a sufficient condition for blocking by considering the vorticity equation and imposing the constraint that the vorticity (potential vorticity) be conserved (Emanuel, 1984). The vorticity equation is simply the curl of our momentum equation, which yields a y-component of vorticity,

$$\frac{d}{dt}\nabla^2\psi = \frac{g}{\theta_o}\frac{\partial\theta'}{\partial x}, \tag{4.2.9}$$

where $u = -\frac{\partial\psi}{\partial z}$ and $w = \frac{\partial\psi}{\partial x}$. The other components of the vorticity are easily shown to be conserved since $\frac{\partial\theta'}{\partial y} = 0$. So we can simply consider the y-component of the vorticity to understand the impact of the step on the flow.

Considering only the y-component of the vorticity, the above equation can be simplified by first noting that if the flow is adiabatic, then

$$\theta' = \theta(\psi), \tag{4.2.10}$$

which means that the flow along a streamline does not change θ'. This implies that the stratification changes only if there is a change in the spacing of the streamlines. Given this relationship, then the horizontal gradient of potential temperature can be written as follows:

$$\frac{\partial\theta'}{\partial x} = \frac{\partial\theta'}{\partial\psi}\frac{\partial\psi}{\partial x} = \frac{d\theta}{d\psi}w = \frac{d\theta}{d\psi}\frac{dz}{dt}, \tag{4.2.11}$$

which makes our vorticity equation,

$$\frac{d}{dt}\left(\nabla^2\psi - \frac{g}{\theta_0}\frac{d\theta}{d\psi}z\right) = 0. \tag{4.2.12}$$

This vorticity equation can be solved for the resultant flow except for an undetermined integration constant. To obtain the solution, first consider the following relationships that occur since θ and ψ are strictly functions of z. The parameter $\frac{g}{\theta_0}\frac{d\theta}{d\psi}$ must be a constant both upstream and downstream as the potential temperature cannot change following a streamline (adiabatic flow). Thus, this parameter can be written as,

$$\frac{g}{\theta_0}\frac{d\theta}{d\psi} = \frac{g}{\theta_0}\frac{d\theta}{dz}\frac{dz}{d\psi} = -\frac{N^2}{u_0}, \tag{4.2.13}$$

where u_0 is the wind for the upstream profile. Also for the upstream profile, the height z can be written as a function of the wind and streamfunction as follows:

$$z = \frac{dz}{d\psi}\psi = -\frac{\psi}{u_0}. \tag{4.2.14}$$

Making use of these relationships and assuming that the streamlines must be flat far enough upstream and downstream from the step, then our equation becomes the following differential equation where the left-hand side represents conditions downstream and the right-hand side, the conditions upstream from the step.

$$\frac{d^2\psi}{dz^2} + \frac{N^2}{u_0}z = -\frac{N^2}{u_0}\frac{\psi}{u_0}. \tag{4.2.15}$$

This relationship must hold downstream, and solutions for this differential equation can be obtained. The equation has solutions of the form,

$$\psi(z) = -u_0 z + u_0 h \cos\left(\frac{N}{u_0}(z-h)\right) + B\sin\left(\frac{N}{u_0}(z-h)\right), \tag{4.2.16}$$

where B is an undetermined integration constant. By differentiating the solution with respect to z, we get the following equation for the wind downstream:

$$-u = \frac{d\psi}{dz} = -u_0 - Nh\sin\left(\frac{N}{u_0}(z-h)\right) + B\frac{N}{u_0}\cos\left(\frac{N}{u_0}(z-h)\right). \tag{4.2.17}$$

To determine B, we must impose some boundary condition on the solution. If we assume that at the level of the step height, the downstream wind should be equal to the upstream wind, then the coefficient B must be equal to zero. This yields the solution for the wind to be as follows:

$$-u = \frac{d\psi}{dz} = -u_0 - Nh\sin\left(\frac{N}{u_0}(z-h)\right). \tag{4.2.18}$$

Upon examination of this solution for the downstream wind, if $Nh > u_0$, then u must reverse sign at some level below the height h. This can be seen by evaluating the sine function at some level $z < h$. The sine term becomes negative and greater than u_o, implying that the sense of the solution changes sign. Under these conditions, our solution is no longer valid for the assumptions used to derive it, which presumably corresponds to the flow being blocked below this level. This condition for the flow to be blocked corresponds to the same condition derived previously for the more restrictive problem of having the pressure perturbation vanish downstream, namely the Froude number $\frac{u_0}{Nh}$ less than 1. So even for this more general problem, where the perturbation pressure does not vanish downstream, we still get the same sufficient condition for flow blocking to occur. Numerical simulations of this situation with a more complete model suggest that indeed the flow becomes stagnant (blocked) below a level

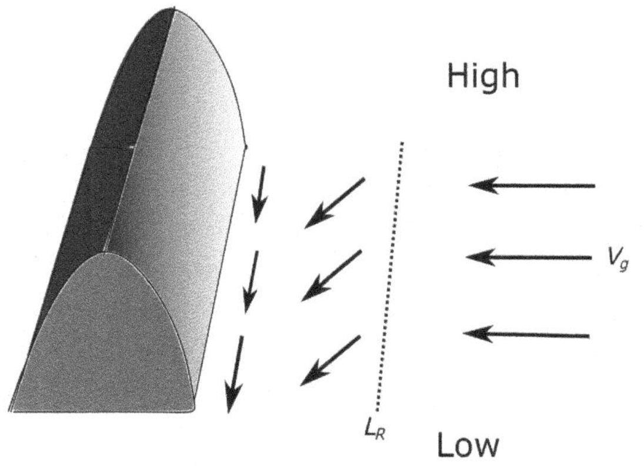

Fig. 4.2 Flow interaction with a 3D topographic barrier. A uniform pressure gradient occurs in the along-barrier direction, and the incident flow is given as geostrophic V_g for this pressure gradient. The dashed line at a distance L_R upstream from the barrier represents the point at which the flow first feels the barrier, which is the Rossby radius. Within this distance, the flow can not adjust geostrophically as the cross-mountain component is blocked. The flow then accelerates in the along-barrier direction down the pressure gradient.

$h - h'$ such that $\frac{Nh'}{u_0}$ is approximately 1. For the two-dimensional problem described here, this stagnation extends indefinitely upstream from the step.

For realistic three-dimensional problems where flow can occur along the mountain direction, the mathematical problem becomes more difficult and can not be easily solved analytically. However, some simple physical reasoning allows us to extend this two-dimensional result to include the Earth's rotation in the problem from a three-dimensional perspective. For this more realistic problem, results indicate that the block extends only a finite distance upstream from the mountain (Fig. 4.2). This occurs due to the additional energy source provided by the Coriolis acceleration as the air becomes subgeostrophic near the mountain. This is evident if we consider the simple example illustrated in Fig. 4.2. The air upstream is initially in geostrophic balance with a uniform north-to-south pressure gradient. As the air approaches the mountain, at some distance, upstream it begins to feel the effects of the mountain and starts to decelerate. This makes the wind speed subgeostrophic, and the air must turn toward lower pressure to the south. As the air begins to accelerate in this along mountain direction, the Coriolis acceleration across the mountain increases, and the air goes up and over the mountain. It can be shown that if the air starts far enough upstream, it will always make it over the mountain due to this Coriolis acceleration. Hence, there exists a finite distance upstream (L) within which the air stagnates, and this distance depends upon the incident flow speed or pressure gradient to become a mountain-induced Rossby radius.

The mountain Rossby radius identifies the upstream distance where the air first becomes subgeostrophic due to the mountain and is given by $L_R = \frac{U}{f}$. For the flow to be blocked (modified by the mountain), the upstream velocity is given by $U = Nh - U_0$ as found in 4.2.18, which is due to the Froude number being equal to one. This indicates that the mountain Rossby radius is given by $L_R = \frac{Nh - U_0}{f}$, which corresponds to the finite upstream distance within which blocking occurs. The upper bound on this length scale is given by the lowest upstream wind that will make it over the mountain. Namely as U_0 tends toward zero, the Rossby radius tends toward its maximum distance, which is bounded by $L_{max} = \frac{Nh}{f}$. Both modeling and observational studies confirm that this length scale does correspond to the maximum upstream radius of influence of the mountain.

This relationship implies that the slope of the topography is important in determining whether blocking will occur, not just the height of the mountain. To see this, consider that the half-width of the mountain must be less than or equal to L_{max}. Thus, $L < \frac{Nh}{f}$ corresponds to a blocked regime. This implies that the mountain slope $\frac{h}{L}$ must be greater than $\frac{f}{N}$ for blocking to occur or less for no blocking to occur. Thus the mountain steepness is directly related to the occurrence of blocking. This relationship implies that steep mountains are more effective at producing blocked flow, and the steepness depends upon the stability and location $\frac{f}{N}$. This scaling factor indicates that for larger values of N, a less steep mountain can cause blocking. Consider some typical ranges of N (10^{-1} to 10^{-2}) and f (10^{-4}), which yield values of $\frac{f}{N}$ from $\frac{1}{1000}$ to $\frac{1}{100}$. Thus slopes steeper than these values will produce blocking. For example, a 1 km high mountain could block the flow 1000 km upstream. Note that these values are based upon the upper bound on the upstream distance, which is based upon the wind speed tending toward zero. For winds greater than this minimum speed, the upstream distance decreases ($L_R = \frac{Nh - U_0}{f}$). So this distance can be shortened very quickly by increasing wind speed. To see this, consider a 1 km high mountain and N equal to 10^{-2}, which implies a mountain slope greater than $\frac{1}{100}$ for flow blocking. If the incident wind speed is 5 m/s, then the distance upstream is 50 km, but if the incident wind speed increases to 8 m/s, the upstream distance for blocking drops to 20 km. So mountain slope is very important to yield flow blocking, and for higher wind speeds, steeper topography is required to produce flow blocking.

Note that the arguments presented in this section apply only to the component of the wind that is perpendicular to the mountain. The wind component parallel to the face of the mountain can still maintain geostrophic balance and will be unaffected by the mountain. Even a 50 m/s low-level jet can be blocked if it is primarily flowing parallel to the topography.

The basic theory is based on uniform stratification and uniform incident flow. However, both the stratification and incident wind typically have vertical structure in the layer below the top of the topography. The impact of this vertical structure can be deduced by considering the mountain Rossby radius ($L_R = \frac{Nh - U_0}{f}$), where N and U_0 are allowed to vary in the vertical. For static stability increasing with height, the effective Rossby radius increases with height, which steepens the upstream face of the blocked region. Hence, a more vertical boundary can occur when compared to the wedge shape found under uniform stratification. The opposite occurs if the static stability decreases with height, such that a shallow, flat blocked region is observed. Vertical shear in the incident flow can produce similar effects as the Rossby radius changes for different flow speeds. For winds increasing with height, the upper portion of the flow has a shorter Rossby radius, which results in a shallow, flat blocked region. If the winds decrease in the vertical, then the upper part of the flow has a longer Rossby radius and the block becomes more vertical. Also note that this applies to the cross-mountain component, and so speed shear can be produced by directional shear in the total flow.

4.3 Mountain Waves and Flow over Mountains

The other response of flow interaction with topography occurs when the Froude number becomes greater than one. This implies that the winds are strong enough for the given stratification to easily make it up over the mountain. A portion of the upstream flow may still be blocked, but the upper portion of the flow gets forced up and over the topography to result in the same kind of response. Let's consider what happens in this case by considering a stably stratified flow over a mountain, such that no air goes around the mountain (infinite mountain range). In this case, the air is forced up the mountain on the incident side by mechanical lifting and down the other side by negative buoyancy. To see this, consider the example illustrated in Fig. 4.3. On the windward or ascending side, the air must decelerate as energy is used to lift the parcel up the side of the mountain. The wind speed decreases on the upstream side of the

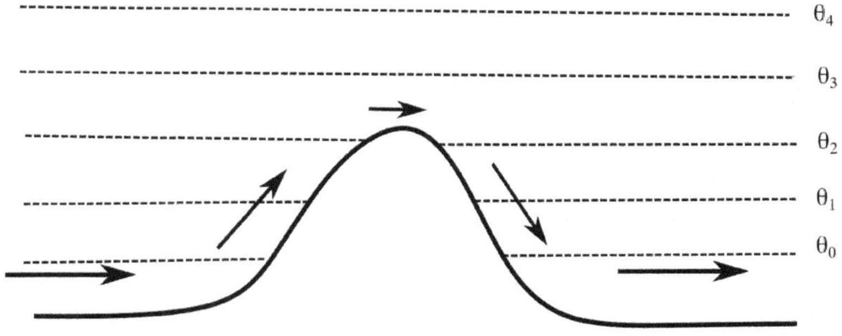

Flow over a hill with background stable stratification.

Fig. 4.3

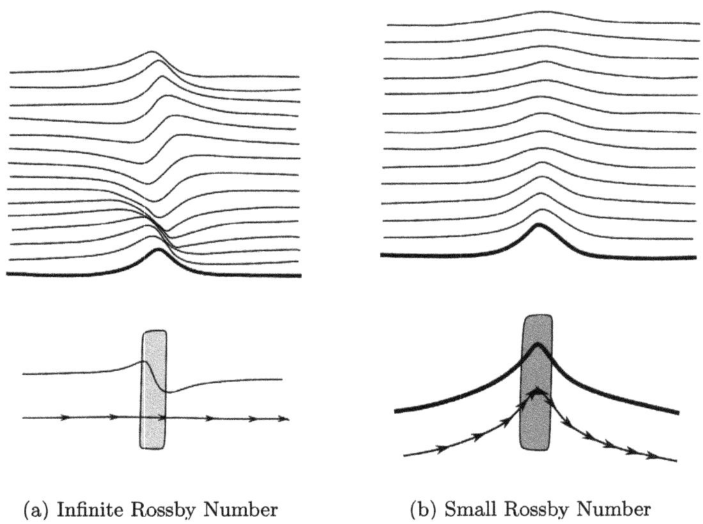

Fig. 4.4 Flow response to a 2D topographic bump under different Rossby numbers. Top diagram shows the isentropes in a vertical cross section across the barrier. The bottom part of the diagram shows a horizontal depiction of a representative pressure contour (solid line) and steamline (arrows) across the barrier (shaded rectangle).

mountain. This air parcel conserves its potential temperature (dry adiabatic motion) and finds itself at the top of the mountain with a potential temperature colder than the surrounding air. On the lee or descending side, the air at the top of the mountain must accelerate due to the gravitational acceleration of this parcel that is colder than its surroundings. This negative buoyancy brings the parcel back down the slope, and the wind speed increases back toward its upstream value. The downward acceleration results in overshooting the equilibrium level and excites a buoyancy oscillation. This buoyancy oscillation is essentially the process by which vertically propagating mountain waves are initiated. The actual response of a flow across a two-dimensional mountain barrier depends upon a variety of factors, including the width of the mountain, stratification distribution, and incident wind speed. Pierrehumbert (1986) shows that for uniform stratification, the nature of the topographic response depends primarily on the speed of the flow relative to the size of the mountain. The mountain Rossby number ($Ro = \frac{u}{fL_m}$) and the Froude number ($Fr = \frac{u}{Nh_m}$) determine the character of the flow response to the topography.

For a large Rossby number ($Ro \approx \text{inf}$), the wind is either strong and/or the mountain is narrow, and the flow can not adjust geostrophically. Vertically propagating gravity waves perturb the structure above the mountain and lead to a trough in the potential temperature field that tilts upstream with height as shown in Fig. 4.4(a). Lifting on the upstream side of the mountain causes adiabatic cooling, and descent on the downstream side causes adiabatic warming. This forced heating pattern results in a windward pressure ridge and a leeward pressure trough. The wind speed is greatest at the top of the mountain due to the induced cross-mountain pressure gradient of the perturbation pressure field. The flow is essentially not deflected as it crosses the barrier and is seen to be highly ageostrophic. Complexity in the stratification can greatly alter the response and will be examined below.

The dynamic response to flow over a relatively narrow mountain, where the Rossby number is large (no geostrophic response), can be characterized as linear mountain waves. The mathematics are given by the following governing equations:

$$U\frac{\partial u'}{\partial x} + \frac{\partial U}{\partial z}w' + \frac{1}{\rho_0}\frac{\partial p'}{\partial x} = 0 \tag{4.3.1a}$$

$$U\frac{\partial w'}{\partial x} - g\frac{\theta'}{\theta_0} + \frac{1}{\rho_0}\frac{\partial p'}{\partial z} = 0 \tag{4.3.1b}$$

$$\frac{\partial u'}{\partial x} + \frac{\partial w'}{\partial z} = 0 \tag{4.3.1c}$$

$$U\frac{\partial \theta'}{\partial x} + \frac{N^2\theta_0}{g}w' = 0 \tag{4.3.1d}$$

which represent the simplest form of the governing perturbation equations for two-dimensional flow over topography. The first two equations represent the horizontal and vertical momentum equations. The third equation is the continuity equation, and the fourth equation is the thermodynamic energy equation. This set of equations can be combined into a single equation for the perturbation vertical velocity, which is Scorer's equation (Scorer, 1949).

$$\frac{\partial^2 w'}{\partial x^2} + \frac{\partial^2 w'}{\partial z^2} + l^2(z)w' = 0, \tag{4.3.2}$$

where $l^2(z)$ is the Boussinesq form of the Scorer parameter (Scorer, 1949) and is given by

$$l^2(z) = \frac{N^2}{U^2} - \frac{1}{U}\frac{\partial^2 U}{\partial z^2} \tag{4.3.3}$$

The solutions to this equation depend upon the sign of the parameter l^2. Two basic solutions exist for this equation, one when l^2 is greater than k^2 (forced wavenumber) and one when l^2 is less than k^2 and are given as follows. k^2 is the wavenumber associated with the topography, which represents the wavenumber of the forcing.

$$w(x,z) = -Uh_m k e^{-(k^2-l^2)^{\frac{1}{2}}z} sin(kx) \qquad k^2 - l^2 > 0 \tag{4.3.4a}$$

$$w(x,z) = -Uh_m k sin(kx + mz) \qquad k^2 - l^2 < 0. \tag{4.3.4b}$$

The sign of this parameter and its vertical distribution lead to a variety of wave responses. The wave response can be damped when the $k^2 - l^2$ is positive, which produces evanescent waves that decay above the mountain as

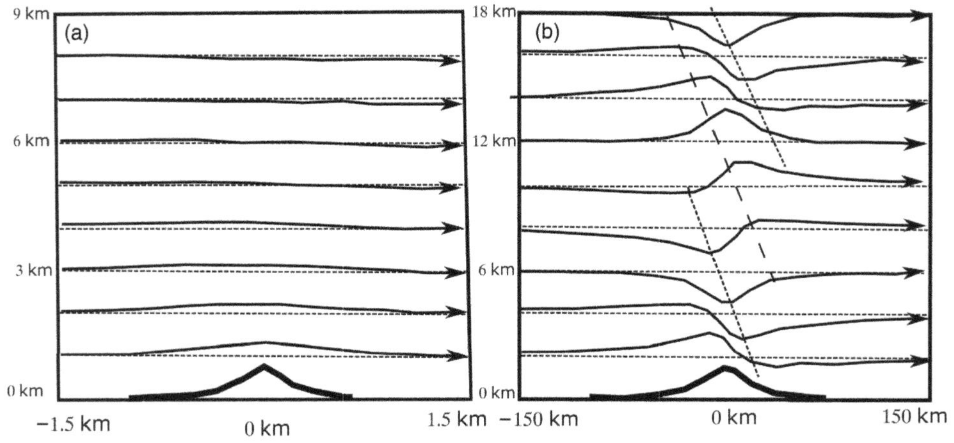

Fig. 4.5 Wave response when (a) $k^2 - l^2 < 0$ producing evanescent waves and (b) $k^2 - l^2 > 0$ producing vertically propagating waves. The mountain width varies over two orders of magnitude from (a) 1 km to (b) 100 km. After Durran (1986).

shown in Fig. 4.5(a). This condition represents situations when the wavelength of the topography is shorter than the wavelength of a free oscillation in the atmosphere. For negative values of $k^2 - l^2$, vertically propagating waves are produced that tilt upstream in the vertical to result in perturbation low pressure on the lee side and high pressure on the windward side, as shown in Fig. 4.5(b). This condition represents situations where the wavelength of the topography is larger than the wavelength of a free oscillation in the atmosphere. Hence as depicted in Fig. 4.5, evanescent waves occur with narrow mountains (1 km wide in Fig. 4.5(a)) and vertically propagating waves occur with broad mountains (100 km wide in Fig. 4.5(b)) for the same atmospheric stability and flow speeds. The nature of the mountain wave response depends on details of the vertical wind profile, stability profile, width of the mountain, presence of moisture, and a variety of other factors. The basic response is a vertically propagating gravity wave as seen in equation 4.3.4b and shown in Fig. 4.5(b). In this case, the wind and static stability profile are uniform in the vertical.

A trapped wave response can occur if the wind speed increases and static stability decreases above the mountain-top level. This results in the Scorer parameter decreasing upward, such that the upper layer may become evanescent and cause the wave to refract back down into the lower layer. This wave refraction from the upper layer (where winds are 1.6–2 times as strong) back into the lower layer causes a constructive interference pattern to be set up, so that many wave crests can be observed well downstream from the mountain. Trapped waves are often seen in satellite imagery as multiple wave clouds extending downstream from the topography as shown in Fig. 4.6, where cloud bands extend across Eastern Oregon. Trapping can be produced by wind speed increases with height or stability decreases or both. Typically the wind speed must increase by 1.6–2 times

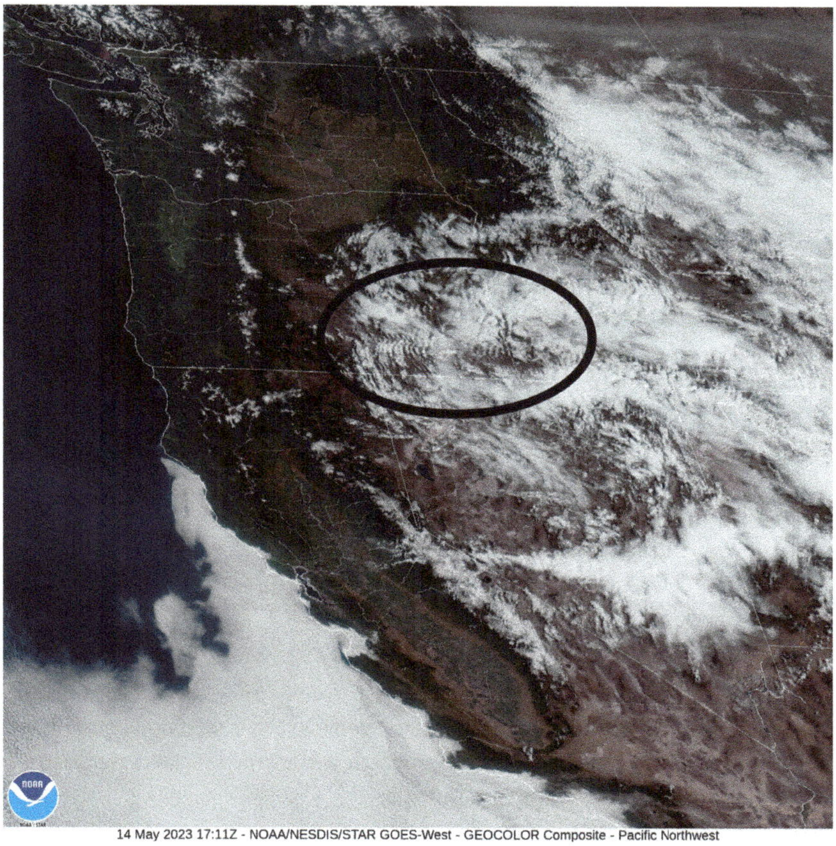

Fig. 4.6 Satellite image from May 14, 2023, showing trapped mountain waves over the southern Oregon (marked by oval).

within about 200 hPa above the mountaintop for trapping to occur. Static stability decreases alone are often insufficient to result in trapping.

When the cross-mountain flow direction (or speed) changes to produce no cross-barrier flow, a mean state critical layer is produced that prevents waves from propagating through this layer, and all of the wave energy is confined to the lower portion of the atmosphere. Reverse shear (cross-mountain winds decreasing upward) can also prevent vertical energy propagation by producing a self-induced critical layer (where the flow locally reverses direction). Either a mean state critical layer or a self-induced critical layer (reverse shear profile) forces large downward energy propagation in the lee. This can set up a strong downslope windstorm response on the lee slope as shown in Fig. 4.7. In this situation, the mean state critical layer prevents any wave growth above and creates only the wave trough below the mountain on the lee slope. Hence all the flow is channeled downslope to result in a windstorm with speeds more than double the incident flow speeds.

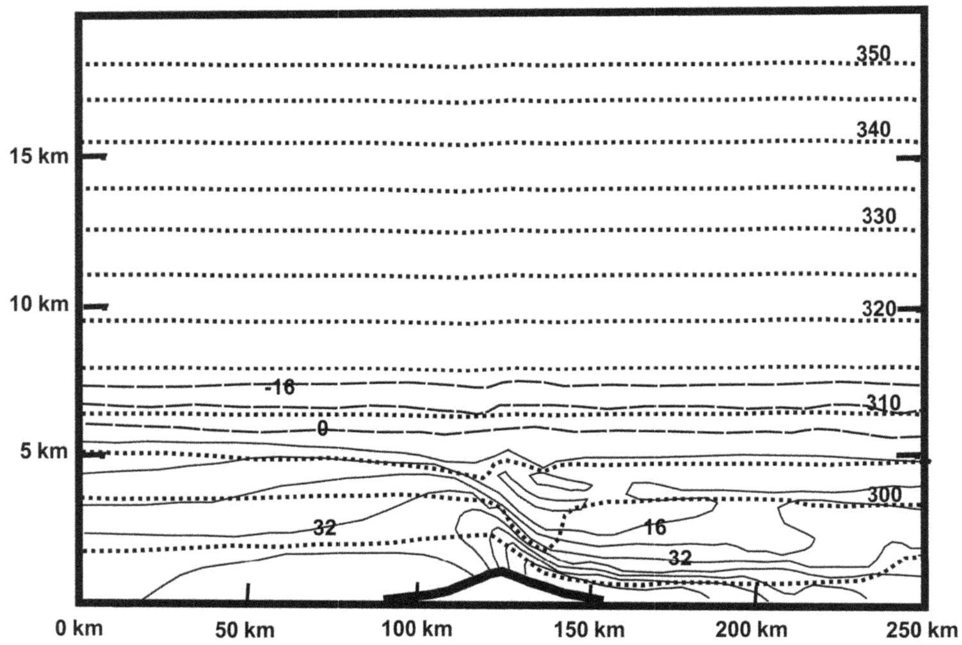

Fig. 4.7 Mountain wave response with a mean state critical level to produce a downslope windstorm. Dotted lines are isentropes, and thin solid and dashed contours are the isotachs of the cross-mountain flow. Isotach contours are every 8 kts, and positive values indicate flow from the left (downstream) and negative values indicate flow from the right (upstream).

The wave response to changes in the mean flow and stability is a complicated problem and can produce a variety of responses. The reader is referred to Durran (1986) for a more complete treatment of linear mountain waves and their responses to varying background flow conditions. For relatively uniform wind shear, the static stability tends to determine the amplitude of the mountain wave response, where higher stability produces a higher amplitude wave response. This sets up a larger lee trough/windward ridge amplitude and increased flow across the topography. The amplitude of the lee trough can be important when considering impacts in the coastal region and how they may vary along the coast.

For small Rossby numbers ($Ro \approx 0.01$), the wind is either weak and/or the mountain is broad, which allows the flow to adjust geostrophically as it crosses the barrier. In this case, as the air is lifted on the upwind side, the adiabatic cooling produces higher surface/sea-level pressure, which turns the air parcel to the left around the high pressure as shown in Fig. 4.4b. The air flow toward the barrier decreases as we get closer to the mountain. The streamlines are deflected toward the left as parcels approach the barrier. At the top of the mountain, the air is coldest and the induced pressure is the highest. This pressure ridge is centered over the barrier and decreases on the lee slope. Flow at the top of the barrier is weakest and then accelerates

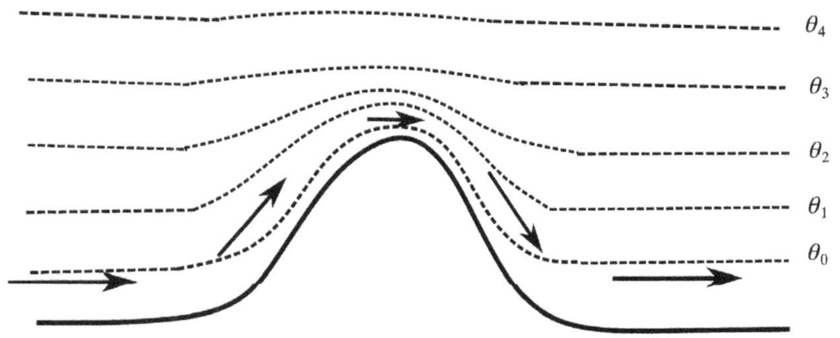

Fig. 4.8

Flow over a hill after adiabatic adjustment. The surface flow has a potential temperature of θ_0, which must conform to the surface as the air flows over the hill and conserves its potential temperature.

on the lee slope as the adiabatic warming lowers pressure and the flow turns back to the right. The pressure and temperature perturbation are maxima near the surface and damped above the mountain. After a continous flow of this type, the potential temperature distribution takes on the shape of the mountain for a dry adiabatic flow as shown in Fig. 4.8. Given some level above the top of the mountain where the effect of the mountain is minimal (potential temperature contours stay flat), the distribution of potential temperature might look like that shown in Fig. 4.8.

In summary then, for high Froude number flows, lee troughing and windward ridging effects are the primary model of flow interaction with topography when the flow is relatively strong, the stratification is not strong enough to result in blocking, and the mountains are narrow. Details of the structure of this interaction depend upon the exact conditions within the atmosphere, and specific examples will be considered in the Chapters 8 and 11.

4.4 Thermal Impacts of Mountains

The impact of mountains on the flow is not strictly mechanical as thermodynamic processes are also important. As noted in Section 4.3, the thermal structure around mountains is often the adiabatic response to flow over the mountain and the type of mountain wave that is produced. This structure is however modified by diabatic processes produced by surface heating differences that occur due to slope variations. Under weak flow conditions, surface heating and cooling occur on the mountain slope sooner than over the adjacent flat areas. This is illustrated in Fig. 4.9, which shows the flow response in the Salinas Valley in California. The flow is primarily onshore from the Northwest, along the axis of the flat valley. However,

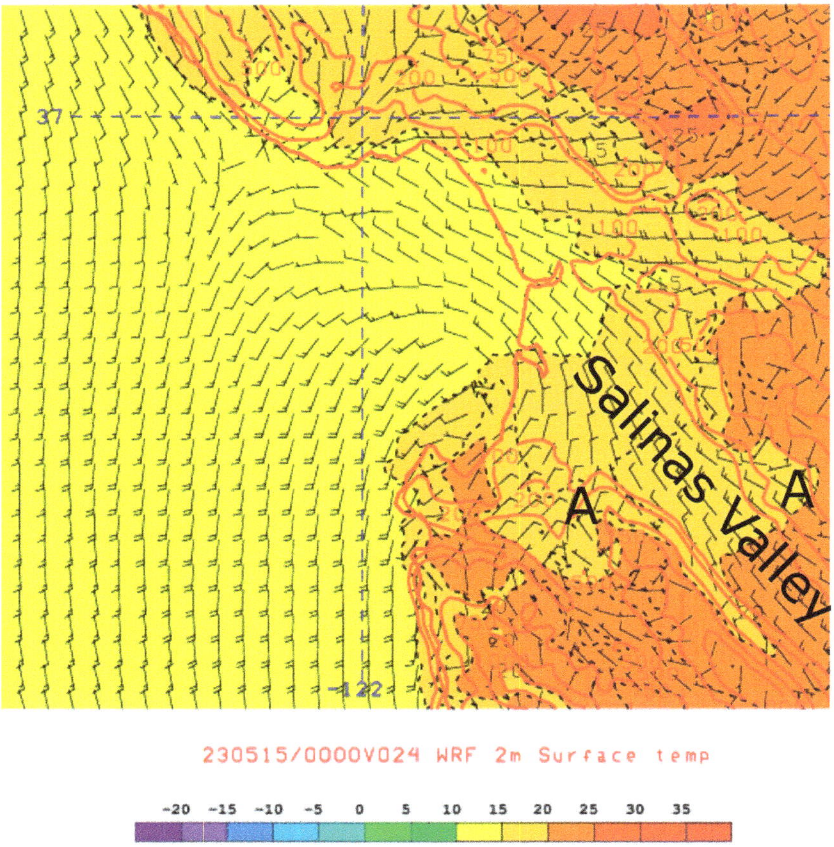

Fig. 4.9 Thermally driven circulation in topography in model simulation near the Monterey Bay of California. The color shading is 2 m air temperature, red contours are terrain elevation, and black wind barbs are 10 m winds. Higher elevations and inland areas are warm. Flow is northwesterly across the coast into the Salinas valley due to thermal forcing across the coast during the late afternoon. The flow goes upslope along the edges of the Salinas valley, highlighted by the arrows, toward the warmer temperatures at higher elevations.

strong slope heating has occurred on the adjacent slopes to cause the flow to turn upslope along the edges of the valley. This is particularly evident at the locations marked A. The along-slope thermal gradient that can set up downslope flows during cooling (katabatic or drainage flows) and upslope flows during warming (mountain-valley flows), as illustrated for the Salinas Valley in Fig. 4.9.

Under stronger flows where mountain waves may occur, the slope heating and cooling alter the low-level static stability, which can impact the wave response near the surface. Stronger near-surface mixing due to heating on the lee slope can promote greater downslope wind speeds by mixing the momentum down to the surface. Nocturnal cooling may weaken the near-surface winds by producing greater near surface stratification to inhibit mixing. These thermal effects lead to diurnal modulation of downslope windstorms.

4.5 Exercises

4.1 Flow blocking is presumed to occur when the Froude number is less than 1. What determines the upstream distance where blocking effects are no longer important? Write down the formula for determining this distance, assuming the mountain modified flow is given by $Nh - U_0$. Using this formula, determine the range of distances for a 1 km high mountain for Brunt-Vaisala frequencies between 0.008 and 0.02 s^{-1} and incident flow speeds between 2 and 10 m/s.

4.2 The vertical structure of the blocked flow typically looks like a wedge of cooler air trapped against the mountain. Sketch this structure and describe why the air near the mountain should be cooler. Consider easterly and westerly flows impinging on the barrier, which is more favorable to maintain a block (consider where the air in the blocked region comes from) and why? If there is strong vertical shear in the incident flow, how will this modify the shape of the wedge?

4.3 For high Froude number flows, mountain wave responses are expected when the width of the mountains is small enough not to allow geostrophic adjustment (mountain width is less than the Rossby number) but not too small. If the mountain width is too small, the mountain waves tend to be damped (evanescent). Calculate the width at which mountain waves produced by a 15 m/s flow with a Brunt-Vaisala frequency of 0.01 will be damped. If the flow speed increases to 30 m/s, does the minimum mountain width increase or decrease?

4.4 Both static stability changes and wind speed changes in the vertical can lead to a trapped wave response when the upper layer becomes evanescent.

- Assume that the Brunt Vaisala frequency is 0.01 and the cross-mountain flow is 15 m/s for the layer interacting with the mountain. Assess the wind speed change that must occur in the vertical to trap waves forced by a 15 km mountain width (wavelength).
- Assuming a 30 m/s constant flow profile but changing the Brunt Vaisala frequency in the vertical, what must the Brunt Vaisala frequency in the lower layer be to give trapping if the upper layer is 0.005 and the mountain is 15 km wide? Is the upper layer evanescent?

4.5 If we consider the situation along the California coast during the summer, it is reasonable to expect that the flow will be blocked given the strong low-level inversion. Calculate for wind speeds of 2, 5, and 10 ms^{-1} the Froude number and upstream blocking radius for the California coast. Assume that the height of the mountains is 1000 m, and that the Brunt-Vaisala frequency can be determined using a two-layer approximation ($N^2 = \frac{g}{\theta_o}\frac{\Delta\theta}{h}$), where h is the height of the inversion and $\Delta\theta$ the temperature change across it. (assume a typical summertime inversion of about 12°C at a height of 500 m).

4.6 Consider air freely flowing over a mountain that is 1 km high. Determine the magnitude of the pressure perturbation due to the air flow over the mountain, assuming that the background stratification is given by the standard atmosphere lapse rate and that the mean thermal change in the surface to mountain top layer is about 1/2 the temperature difference a parcel would have after being lifted to mountain top. Assume that the air prior to being forced up the mountain has a pressure of 1013 hPa and temperature of 288°C.

4.7 For mountain wave regimes, the flow response is determined by a number of factors, including the mountain shape and width as well as the incident flow structure. Describe how the following factors influence the response (type of waves expected) and why?

- Mountain width
- Wind profile
- Stability profile

4.8 Many mountain wave studies assume symmetric background conditions (same up and downstream), but the atmosphere often produces very different structures up and downstream. Describe what impact a low-level stability increase downstream from a mountain might have on the nature of the wave response.

4.9 The condition for flow blocking to occur is for the Froude Number ($\frac{U}{Nh}$) to be equal to unity. Derive an expression to show how the Froude number varies vertically with respect to height below the mountaintop for a uniform wind speed profile. Following this idea, derive an expression for the vertical variation of the blocking radius given a wind profile that varies in height. Show how the slope of the blocked region depends on wind speed for a uniform wind profile. Show how this will vary for a linear wind speed increase with height. Utilizing the expressions you derived, show whether shear increases or decreases the slope of the blocked region.

4.10 Consider the situation inside the blocking Rossby radius and describe why the wind turns to be parallel to the mountain barrier. Utilize a simple form of the horizontal momentum equation (perturbation form) where friction and coriolis terms are ignored to derive an expression for the wind speed change in the along-mountain direction. Based on this relationship, how fast would the wind be after traveling 100 km along the barrier, assuming a cross-barrier geostrophic flow of 10 ms^{-1}? Describe what determines the strength of a barrier jet.

4.11 Mountain waves are the response to flow over topography in a stably stratified atmosphere. Utilizing a Brunt-Vaisala frequency of 10^{-1} and 10^{-2} with a cross-mountain flow speed of 10 m/s, determine the mountain widths over which vertically propagating undamped mountain waves occur for the two stabilities. Note that the wavenumber is the inverse of 2π times the wavelength. Comment on how these results change if you double the wind speed.

4.12 Mountain waves can not propagate through a mean state critical level (zero cross-barrier flow) because the wave equation solution goes to zero at that level (waves can not exist). Comment on what impact this has on the wave response below this level. If this level is 5 km above the top of the mountain or 1 km above the top of the mountain, do you think the surface winds will be different, and why?

5 Sea–Land Breeze Circulations

Sea breeze circulations are one of the most recognized types of coastal weather phenomena. They are ubiquitous to almost any coastal environment and have impacts on local winds, temperatures, air pollution, clouds as well as near-shore ocean conditions. The sea breeze circulation has been recognized since antiquity and ranks as one of the most studied atmospheric phenomena. While the dynamics are very well understood, the details of sea breeze circulations are highly varied and surprisingly sensitive to relatively small factors. We will examine these factors and their impacts on the basic circulation.

5.1 Description of Sea Breeze Circulations

To begin to consider the practical aspects of idealized thermally driven circulations noted in Chapter 3, we will examine the well-known sea/land breeze circulation. The sea/land breeze circulation has had considerable scientific study and is a highly recognized coastal circulation for even the casual meteorologist. Much of the scientific study of the sea breeze in the US has concentrated on the Gulf of Mexico coast of Texas and the coasts of Florida. These relatively flat coastlines simplify the problem to allow one to concentrate on the purely thermally driven aspects of the circulation. Although many studies have focused on these areas, sea breeze circulations are prevalent along most coastlines when synoptic conditions allow their existence.

To begin to develop a conceptual model of the sea breeze, it is helpful to consider the physical processes active at the coastline that force the sea breeze. From there, we can develop a simple mathematical model to better understand the basic circulation as well as start to explore modifying effects on the sea breeze. So for purposes of considering the basic forcing of the sea breeze, let's consider the simplified situation of a flat coastal boundary that runs indefinitely in the along-shore direction with no along-shore variations in properties. This simplified coastline will allow us to develop a simple mathematical model of the circulation in Section 5.2.

The basic forcing mechanism driving sea/land breeze circulations is the development of a coastal thermal gradient to which the flow responds. If

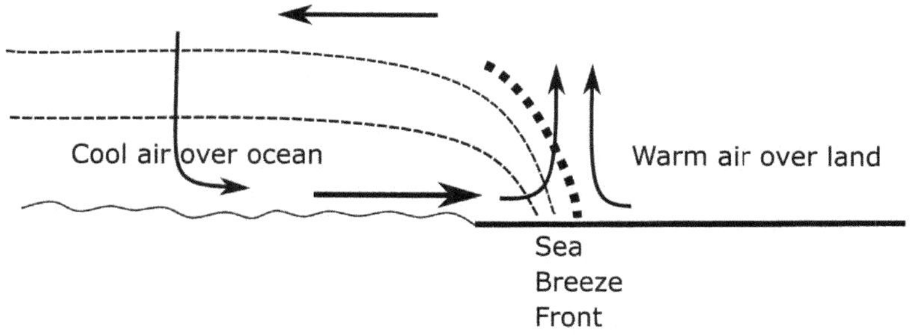

Fig. 5.1 Sea breeze circulation. Thin dashed lines are isentropes showing cool air over the ocean. Thick dashed line is the sea breeze front. Arrows are the vertical and horizontal flow, showing rising motion over the land, sinking motion over the ocean, onshore flow at the surface, and offshore flow aloft.

we consider initially uniform atmospheric characteristics across the coastal boundary and then allow solar radiation to impinge upon the coastline, a thermal gradient develops due to the differential radiative properties and heat capacities of the water and land surfaces. The land surface absorbs solar radiation much more efficiently than the water and the land heats up relative to the adjacent water due to its lower heat capacity. The surface heating is transported into the overlying atmosphere through surface heat fluxes to result in a temperature rise over the land area that does not occur over the water. After some period of heating, a well-developed coastal temperature gradient is present. This sets up a circulation across the coast as depicted in Fig. 5.1, where air rises over the land, flows seaward aloft, sinks over the water, and then flows onshore near the surface to form the sea breeze. The coastal thermal gradient gets concentrated into a sea breeze front as low-level convergence gets established by the onshore flow from the water and stagnant (or weak offshore) flow over the land. The strength of the sea breeze front depends upon the cross-coast thermal gradient and the magnitude of the low-level convergence that causes frontogenesis. The sea breeze front then propagates inland as a gravity current set up by the cool ocean air and onshore winds. At night, the now much warmer land surface radiates more energy in outgoing long-wave radiation to cool the land surface relative to the adjacent water. In the classic sea breeze, this cooling is sufficient to reverse the coastal temperature gradient and the sense of the associated circulation as depicted in Fig. 5.2. While the surface temperatures may go through a relatively symmetric cycle of heating and cooling, the vertical structure can be quite different in the land breeze due to boundary layer stabilization caused by the surface cooling. This results in differences in the structure and evolution of the land breeze compared to the sea breeze. As described, the low-level temperature gradient and winds are observed to undergo a diurnal cycle that follows the diurnal cycle in the surface energy budget forced by the differences in radiative properties.

The sea breeze cycle along the Texas coast has been studied extensively by Hsu (1970) and serves as a useful description of the basic circulation.

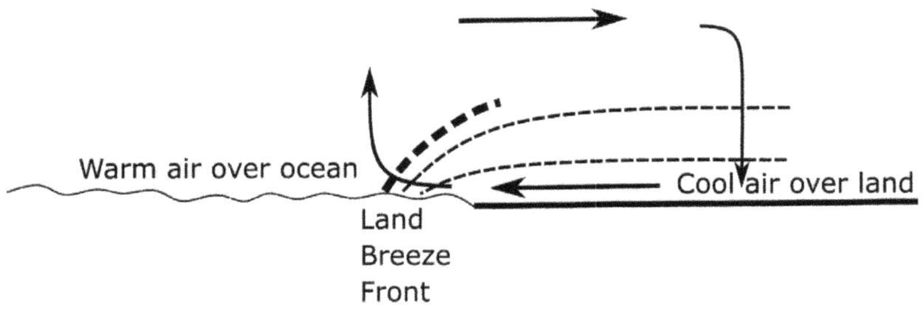

Fig. 5.2 Land breeze circulation. Thin dashed lines are isentropes showing cool air over the land. Thick dashed line is the land breeze front. Arrows are the vertical and horizontal flow, showing rising motion over the ocean, sinking motion over the land, offshore flow at the surface, and onshore flow aloft.

Figure 5.3 adapted from Hsu (2013) shows the typical cycle of winds in a vertical cross-section across the coastline through a complete diurnal cycle from 6:00AM local time to 12:00AM local time at 6-hour intervals. The first schematic (Fig. 5.3(a)) shows the early morning conditions at 6:00AM when the land is cooler than the adjacent ocean and the sun has not yet risen to start the heating cycle. At this initial time, the circulation cross-section shows an offshore-directed flow at the surface with an onshore-directed flow in the upper layer. The circulation extends approximately 30–40 km inland and just off the coast. This is a land breeze circulation due to its offshore flow direction at the surface. The thermal structure has a pool of cool air over the land that has been advected offshore as a land breeze front. As the heating begins and progresses through the morning, the land heats up relative to the ocean so that by 12:00PM (noon), the land is much warmer than the ocean. Rising motion in the warm air coupled with the across-coast thermal gradient initiates onshore low-level flow as shown in Fig. 5.3(b). Over time, the result is the establishment of an onshore directed flow across the coast at the surface with an offshore-directed flow in the 900–700 hPa layer. This is the so-called sea breeze, which is initially limited to about 20 km in either direction across the coast. The sea breeze front develops at the leading edge of this circulation, where the thermal gradient is strong and the wind changes direction from onshore in the cool air to near zero or offshore in the warm air. The wind direction changes result in convergence and the concentration of the thermal gradient into a front. The land breeze observed 6 hours earlier has been pushed inland by the sea breeze circulation, which actually is the result of the interior branches of the land breeze not yet responding to the coastal forcing or being spun down by friction. The convergence between these circulations at the sea breeze front over the land forces upward motion and may result in the formation of cumulus clouds along the sea breeze front. The next 6 hours show the growth and initial decay of the sea breeze as the temperature gradient reaches a maximum

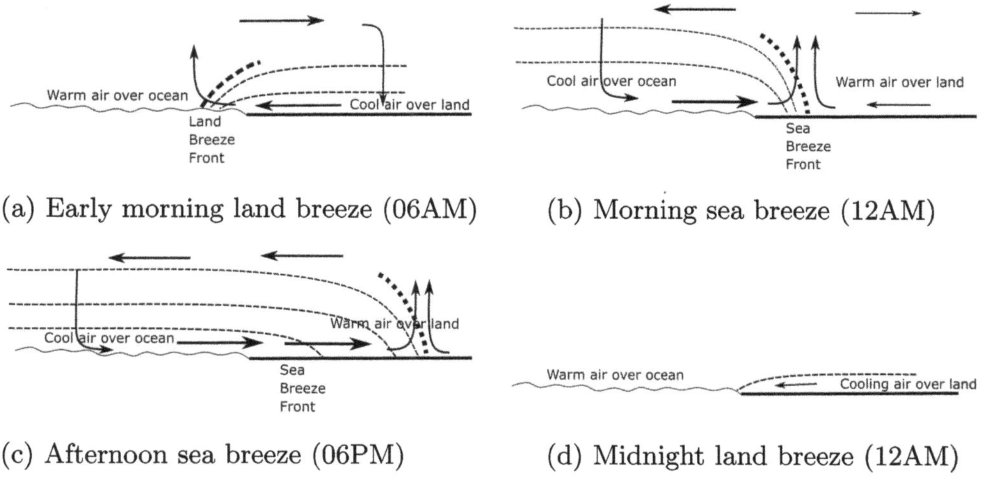

Fig. 5.3 Idealized lifecycle of sea/land breeze evolution. (a) Early morning circulation shows offshore surface flow, rising motion over the ocean, sinking over land, and a land breeze front just offshore. (b) Morning circulation shows onshore surface flow, rising motion over land, and sea breeze front near the coast. (c) Afternoon circulation shows onshore surface flow over a greater cross-coast distance, rising motion further inland, and sea breeze front well inland from the coast. (d) Midnight circulation shows little thermal gradient with cooling over land, very weak offshore surface flow and no vertical motion.

around 3:00PM (not shown) and the sea breeze front moves inland. Figure 5.3(c) shows that by 06:00PM, the sea breeze weakens somewhat as the cool air that has been advected inland is warmed by the underlying land surface. The thermal gradient, while strongest at the sea breeze front, has weakened due to this heating of the cool air. This process continues until sundown and the land is no longer warming. Consequently, the strongest sea breeze is observed around 3:00PM with a cross-coast horizontal extent of 30–40 km in both directions and maximum inland penetration of the sea breeze front. As the thermal gradient begins to weaken due to heating of marine air across the coastline, the sea breeze persists but decreases in both intensity and horizontal extent. In addition, the inland movement of the sea breeze front stops as the cross-front thermal gradient weakens and the depth of the leading edge of the cool air tends to shallow. After sundown, the land cools substantially so that little thermal gradient is present at 12:00AM (midnight) and little or no circulation is evident as depicted in Fig. 5.3(d). During the next 6 hours, the offshore-directed land breeze develops and intensifies as the land continues to cool. This cycle is a classic sea/land breeze cycle, which can occur along many coastlines.

While the evolution of the sea/land breeze cycle described above is typical, the actual evolution on any given day can look quite different. Some key questions that the coastal meteorologist may ask are what are the modifying factors to this classic sea breeze cycle, and what are the intensity of the sea breeze winds in a given situation. To consider the first question, we can think of factors that may directly effect the diurnal heating cycle. These factors include but are not limited to

- time of year,
- type of land surface,
- latitude,
- cloud cover,
- depth of planetary boundary layer (PBL),
- other factors that alter the surface energy balance.

The above factors directly effect the surface energy balance to modify the degree of diurnal heating and temperature change that may be observed in a given situation.

The first four factors relate to the ground absorption of solar energy and the outgoing infrared radiation to determine the surface heating. The time of year relates to the duration of incoming solar radiation and how much the land heats up. For example, summer with long periods of sunshine would favor sea breezes as the land will generally experience considerable warming compared to winter periods. The type of land surface plays a big role in determining the amount of solar radiation absorbed by the ground. Vegetated surfaces absorb less solar radiation than bare ground. Urban environments also tend to absorb more solar energy and heat up more than surrounding vegetated areas. These differences in surface heating due to land use variations are illustrated in Fig. 3.2. Latitude influences the sea breeze circulation in several ways. First, the length of day is essentially a function of latitude, which limits the amount of available solar radiation. Furthermore, the incident angle for solar radiation is determined by latitude, with low incidence angles for higher latitudes. Lower incidence angles reduce the amount of solar radiation reaching the ground. Thus, even at the equinoxes, when the length of day is uniform across latitudes, higher latitudes warm less because of the reduced solar incidence angle. Cloud cover directly impacts incoming solar radiation by reflecting some back to space and reducing the surface warming. However, the impact on surface warming is also partially offset by the downward infrared radiation from the clouds that can contribute to warming. The relationship between clouds and surface warming is complicated and depends on factors about the makeup of the clouds, their level in the atmosphere, as well as their spatial coverage.

The fifth factor, depth of the PBL, has to do with the amount of temperature change that might be realized in a given situation. A deep boundary layer with lots of vertical mixing will have a smaller rise in the surface temperature for a given amount of heating than a shallower boundary layer where the heating all occurs in a thin layer. As we will see in Section 5.3.2, the depth of the heated layer can also impact the magnitude of the sea breeze beyond simply the observed surface temperature rise. All of these factors and others that can be thought of are typically complex functions in space, so that the sea breeze can have considerable along-shore variability for a given situation. To illustrate this variability, consider the diurnal temperature maximum over a small section of the Florida coast that is shown in Fig. 3.2. While solar radiation is relatively constant over the region, the temperature maximum varies from about 26°C (light green shading) in the northern region to over 30°C (yellow shading) to the south along the western coast. These differences primarily depend on what the underlying surface is like. These modifying effects will be explored more completely in Section 5.3.

Prior to considering the development of a simple mathematical model, it is enlightening to consider the basic forcing mechanisms and the processes that will typically dominate in the set of equations. Consider first the circulation equation 3.2.10 that was derived from the horizontal equation of motion in Chapter 3. This must account for the observed development of horizontal winds. The differential heating across the coast produces a horizontal temperature gradient as reasoned above. This temperature gradient will be distributed vertically through much of the boundary layer and consequently produce a pressure gradient that is coincident with the thermal gradient. The warm air over the land area tends to lower the surface pressure relative to the unchanged or cooler boundary layer air over the water. Thus an onshore-directed pressure gradient force arises to which the air must respond by accelerating toward the coastline. From the standpoint of the horizontal equation of motion, the pressure gradient term is the dominant factor. The other terms are the Coriolis force and the friction, both of which act on the air once it begins to accelerate. Is the Coriolis force important in the evolution of the sea breeze? In many situations, the time scale of the diurnal thermal forcing is short compared to the inertial cycle, and we can ignore the Coriolis forcing. For higher latitudes where the inertial period is short, this may not be true. Another situation in which the Coriolis term becomes important is when the pressure perturbation may persist through the nighttime hours. In this situation, a modified background flow is established that is local to the coast and has a strong geostrophic response. The vertical branches of the circulation can be deduced through continuity considerations, which also require the presence of a return flow aloft. Thus the basic structure of the lower atmosphere during a sea breeze circulation can be diagrammed as shown in Fig. 5.1. This basic structure can vary considerably, but for consideration of the basic forcing of this simplified thermally driven circulation, this structure has the essential ingredients.

5.2 Mathematical Model of the Sea Breeze

The closed circulation of the classic sea/land breeze allows us to describe the mathematics governing the circulation using the circulation theorem.

The mathematical circulation is determined by integrating around the closed loop of the circulation. Mathematically, the circulation is defined as

$$C = \oint \mathbf{V} \cdot \mathbf{dl} \approx VL, \tag{5.2.1}$$

where the vector velocity $\mathbf{V}$ is integrated around the loop $\mathbf{dl}$. Approximately, the circulation is given by the product of the mean velocity V, and the total path length of the circulation L. The circulation C is considered to be positive for the land breeze if the positive x direction is taken to be in the onshore direction. The circulation C is then a measure of the intensity of the sea breeze cell, which is related to both the size of the cell and the magnitude of the winds. Using Stokes theorem, we can relate the circulation to the curl of the velocity integrated over the domain of the circulation

$$C = \iint \nabla \times \mathbf{V} dS. \tag{5.2.2}$$

The time rate of change of circulation is given by differentiating equation 5.2.2 with respect to time which results in the following relationship:

$$\frac{dC}{dt} = \frac{d(\iint \nabla \times \mathbf{V} dS)}{dt}, \tag{5.2.3}$$

which allows us to consider the circulation evolution by understanding the change in velocity within the domain.

Using the concepts of circulation identified above, we can develop a simple mathematical model of the perturbation velocity field associated with the sea breeze. To keep things simple, the following things are assumed about the sea breeze circulation:

- The coastline is flat with no along-shore variations in surface properties. This allows us to consider the problem as essentially two-dimensional in the x–z plane.
- The background large-scale flow is constant with respect to time, and the scale of the spatial variations is large compared to the length scale of the sea breeze. This allows us to ignore any interaction with the larger scale in a linear perturbation approach.
- The sea breeze circulation is shallow compared to the scale height of the atmosphere. This allows us to use a vertical equation of motion where the buoyancy is due to potential temperature perturbations only (no pressure perturbations).
- The x-axis points from the ocean toward the land, and the z-axis points vertically. This defines a positive circulation as a land breeze and a negative circulation as a sea breeze.

Armed with these assumptions, we can now derive using the linear perturbation momentum equations, a circulation equation governing the evolution of the sea breeze circulation.

$$\frac{du'}{dt} = -\frac{1}{\rho_0}\frac{\partial p'}{\partial x} + F_x, \tag{5.2.4a}$$

$$\frac{dw'}{dt} = -\frac{1}{\rho_0}\frac{\partial p'}{\partial z} + g\frac{\theta'}{\theta_o}. \tag{5.2.4b}$$

Given these horizontal and vertical momentum equations, we can write them in vector form as

$$\frac{d\mathbf{V}'}{dt} = -\frac{\nabla P'}{\rho_0} + F_x\mathbf{i} + g\frac{\theta'}{\theta_o}\mathbf{k}, \tag{5.2.5}$$

where friction acts only in the horizontal and the buoyancy only in the vertical. The time rate of change of the circulation can be obtained by integrating this equation around a closed loop. The loop is defined to be the cross-shore length scale X and the vertical depth Z_i. The path is defined by $d\mathbf{l} = \mathbf{i}dx + \mathbf{k}dz$ for a total length scale $L = 2X + 2Z_i$.

$$\oint \frac{d\mathbf{V}'}{dt} \cdot d\mathbf{l} = -\oint \frac{\nabla P'}{\rho_0} \cdot d\mathbf{l} + \oint F_x\mathbf{i} \cdot d\mathbf{l} + g\oint \frac{\theta'}{\theta_o}\mathbf{k} \cdot d\mathbf{l}. \tag{5.2.6}$$

Evaluating the integrals, we get the following equation:

$$\frac{d\mathbf{C}'}{dt} = (\overline{F}_x(z_i) - \overline{F}_x(0))L + \frac{g}{\theta_o}(\overline{\theta}'(2) - \overline{\theta}'(1))z_i, \tag{5.2.7}$$

where $\overline{F}_x$ represents the average friction along the horizontal path at the surface (0) and at the top of the circulation (z_i) and $\overline{\theta}'$ is the average potential temperature perturbation along the vertical paths over land (2) and over water (1), respectively. L is the total path length of the circulation. If the friction is approximated as $-ku'$ and the circulation as Lu', then the equation can be written as

$$L\frac{du'}{dt} = -ku'L + \frac{g}{\theta_o}\overline{\Delta\theta}'Z_i \tag{5.2.8}$$

dividing through by L and representing the change in the thermal gradient $\overline{\Delta\theta}'$ as $Acos(\Omega t)$ where $t = 0$ is the time of maximum heating, we get the following differential equation:

$$\frac{du'}{dt} = -ku' + Acos(\Omega t). \tag{5.2.9}$$

This equation can be integrated to find the solution to be

$$u'(t) = B\exp^{-kt} + A(k^2 + \Omega^2)^{-1}(\Omega sin(\Omega t) + kcos(\Omega t)), \tag{5.2.10}$$

where B is an undetermined constant of integration. B can be shown to be 0 given that if there is no thermal gradient ($A = 0$), then no circulation should exist and $B = 0$. The solution is then given by

$$u'(t) = \frac{g}{\theta_o}\overline{\Delta\theta}'Z_i(k^2 + \Omega^2)^{-1}(\Omega sin(\Omega t) + kcos(\Omega t)). \tag{5.2.11}$$

Based upon this simple mathematical model of the sea breeze, we can estimate the maximum wind in the sea breeze at time ($t = 0$), the time of maximum thermal gradient to be

$$U_{\max} = \frac{kg\overline{\Delta}\theta' Z_i}{(k^2 + \Omega^2)\theta_o L}, \tag{5.2.12}$$

where the quantities have been defined during the derivation above. Using values typical of the Texas coast, $\Delta\theta' = 5°$, $\theta_o = 300\text{K}$, $k = 2 \times 10^{-5}$, $L = 200$ km, $Z = 3$ km, and $\Omega = 7.3 \times 10^{-5}$, we can estimate the maximum wind to be about 8.5 m/s. This value agrees rather well with the observations of Hsu (1970).

Using our approximate expression for the maximum winds, we can explore the impact of certain factors on the sea breeze intensity. First, as the thermal perturbation (temperature gradient $\Delta\theta$) increases, we expect to see an increase in the sea breeze winds. This is consistent with our model and observed behavior. Next, an increase in friction (k larger) results in an intensity decrease. The length scale and depth of the circulation produce opposing effects. Deep circulations would imply stronger circulations than shallow ones. Large horizontal scales tend to produce weaker winds than short horizontal scales. These scale relationships presume that the circulation is unchanged by factors that may limit scales, which is not always true. For example, strong stability that may inhibit the vertical scale may also act to increase circulation due to an increase in the thermal difference across the coast by limiting heating to a shallow layer.

Aside from these relationships, it is also useful to explore the sensitivity of the winds to the various factors over believable ranges that occur in the atmosphere. If we consider a range of thermal perturbations from 0 to 15°C, then for the other factors being the same as those quoted above, the winds would vary from 0 to 26 m/s. For a depth range of 100–3000 m, the winds range from 0.3 to 8.5 m/s. For length scales of 25–250 km, the winds would range from 36 to 4 m/s. From this, it is evident that the thermal perturbation and horizontal length scale are likely to be the biggest factors effecting the sea breeze intensity. However, it is important to note that horizontal length scales are determined by the length scale over which the thermal gradient occurs and generally do not occur over the broad range indicated above. The 25-km scale circulation would require a very tightly packed thermal gradient, which may be very hard to find in reality.

The linear model gives some insights into what factors control the real intensity of the sea breeze and can be used as a simple predictive model of the maximum winds. The actual sea breeze evolution is much more complex, as the length scale is determined by the inland penetration of the sea breeze front, and the vertical depth becomes a function of the detailed boundary layer evolution of the heating through the vertical column. These effects on the sea breeze are examined in Section 5.3 as modifying influences on the basic circulation.

Prior to considering the modifying influences, it is important to understand the evolution of the sea breeze under unperturbed conditions with uniform stratification, no along-coast variations, and no background flow. The evolution is depicted in Fig. 5.3 from the initial onset, inland penetration of the sea breeze front, and then subsequent decay. At the initial onset, the thermal gradient occurs at the coastline, and air is rising over the land due to the heating. This sets up the low-level onshore flow and frontogenesis to create the sea breeze front. As the circulation continues, the sea breeze front moves inland and the onshore flow intensifies in response to the increasing thermal gradient. A return flow develops aloft, and subsidence occurs over the water to complete the circulation. Note that as the sea breeze front progresses inland, heating over the land behind the front occurs to produce a broad thermal gradient that continues to force onshore flow. Note that this heating behind the sea breeze front weakens the thermal gradient of the front (gravity current), thereby slowing its inland progress. Finally, as heating stops and cooling begins, the sea breeze starts to decay. The thermal gradient at the surface is eliminated, and the low-level onshore flow starts to decay.

The inland penetration of the sea breeze front determines the cross-coast length scale and relates to the intensity of the sea breeze. The movement of the sea breeze front can be characterized by the dynamics of a gravity current, where a pool of cold, dense air flows out into the warmer, less dense air. The propagation speed of a gravity current is given by $C = (g\frac{\delta\theta}{\theta_0}h^{\frac{1}{2}})$, where $\delta\theta$ is the potential temperature difference across the leading edge of the gravity current and h is the depth of the cold pool. Using this approach, we can calculate the speed of movement of the sea breeze front for a 500 m deep layer and a cross-front thermal difference of 5°C to be 5 m/s. Given this propagation speed, the movement over a 6-hour period can be calculated to be approximately 100 km. This is larger than typical inland penetration, but given the time evolution of the cross-front thermal gradient, from 0°C initially, to maybe 5°C at maximum, then decreasing back toward 0° by the end of the cycle, the average $\delta\theta$ is more like 2°C. This yields a much more typical 30 km of inland penetration. The inland movement can also be considered from a frictional spin-down perspective, where the onshore flow in a layer of depth h is spun down to zero velocity by friction. This approach yields a distance of 500 km to spin down the sea breeze. Clearly while important, the frictional spin down is not sufficient to prevent deep inland movement, and the thermal changes are important to slow the propagation of the sea breeze front.

5.3 Modifying Influences of the Sea Breeze

The sea/land breeze circulation is influenced by a variety of factors. These factors include but are not necessarily limited to synoptic-scale background flow, vertical wind shear, stability, clouds, coastline shape, and topography. The nature of the modifications that these factors impose upon the sea breeze can be deduced, although their exact impact in a given situation depends upon complete knowledge of the situation.

5.3.1 Background Flow

The impact of the synoptic-scale background flow has been studied under a variety of situations both observationally and numerically. The numerical study by Estoque (1962) provides useful insight into the basic modifying effects that the larger-scale background winds can have on the sea breeze. Estoque (1962) considers four basic flow directions for the large-scale background flow in a simple numerical model. The flow directions consist of onshore, offshore, flow parallel to the coast with land on the left, and flow parallel to the coast with land on the right. The series of figures from Estoque (1962) provides insight into how the basic sea breeze is likely to be modified by these wind directions. The no wind case (Fig. 5.4(a)) provides the basic sea breeze structure and evolution. The 11:00 local time figure shows the beginning sea breeze with a well-developed thermal perturbation over the land and a relatively strong thermal gradient just inland of the coast. As the circulation intensifies through the day, the 17:00 local circulation shows stronger onshore winds at low levels and the inland movement of the thermal gradient. In fact, the thermal gradient maximizes at about 30-km inland and above the 200-m surface layer, where heating spreads this thermal gradient across the entire 30-km region from the coast to the sea breeze front. A well-defined updraft is evident at the maximum thermal gradient, which along with the thermal gradient, defines the sea breeze front. Note that the sea breeze return flow aloft is generally rather weak and occurs over a much deeper layer than the low-level inflow.

Now considering the offshore flow case (Fig. 5.4(b)), where there is a 5 m/s offshore flow, the basic sea breeze is modified in several important ways. First, the initial sea breeze is barely noticeable in the offshore background flow at 11:00 local. The offshore flow is weakened slightly right at the coast, but the onshore flow of the sea breeze is not evident at this time. If the winds were plotted as perturbations, then a weak onshore sea breeze is indicated. As the heating progresses, a very concentrated thermal gradient develops within 8 km of the coast. This strong thermal gradient is sufficient to drive an onshore flow just inland from the coast and force the frontal updraft. Stronger subsidence offshore is another major difference from the basic circulation. Thus, a shorter horizontal length scale results, as do stronger vertical motions, which fit our simple mathematical model in Section 5.2. The sea breeze front does not penetrate as far inland. The stronger vertical motion patterns suggest that coastal clouds are more likely inland with the greater vertical lift, which might be sufficient to break a relatively strong inversion. Coastal clearing just offshore would also be likely if coastal stratus were present from the previous evening. Other studies have confirmed these basic results and show that offshore flow of 6–11 m/s eliminates the sea breeze while flows between 4 and 8 m/s tend to prevent inland movement of the sea breeze front (Arritt (1993) and others). The range in speeds arises because the background flow intensity to eliminate or impede the sea breeze depends on the thermal gradient magnitude as well, which can vary.

Considering the onshore-directed case (Fig. 5.4(c)), where there is a 5 m/s onshore flow, the basic sea breeze is modified considerably. The coastal thermal gradient at 11:00 local time becomes much more spread out inland from the coast and is generally less intense than the basic sea breeze gradient. The more diffuse thermal gradient produces less acceleration, and so the prevailing onshore winds are only slightly modified. Somewhat stronger low-level onshore winds are evident by 17:00 local, but there is basically no vertical motion in this onshore flow situation. Hence, coastal clouds are not likely to form (if not already present) or to dissipate (if already present) due to the vertical motion forced by the sea breeze. This response is also consistent with other studies that show even minimal onshore flow 2–4 m/s is sufficient to mask any sort of sea breeze signature.

The coast parallel flow (not shown) produces effects that look like the onshore or offshore large–scale flow depending upon the wind direction. With the land to the right of the wind, the resultant sea breeze is similar to the offshore-directed flow with slightly greater inland sea breeze frontal penetration. The large-scale pressure pattern has lower pressure seaward, and so the surface friction tends to produce a weak offshore component in the boundary layer. With the land to the left, the resultant sea breeze is similar to the onshore flow situation, with at least some intensification of the thermal gradient inland from the coast. Weak vertical motion is evident, and a noticeable sea breeze is evident in the low-level winds. The large-scale pressure pattern has lower pressure inland in this case, and so the friction tends to produce a weak onshore flow in the boundary layer. These flow parallel situations resemble the offshore and onshore flow situations more than the no wind basic sea breeze, which suggests that even weak synoptic-scale flows play an important modifying role.

The impacts of the synoptic-scale background on the observed sea breeze are illustrated in Fig. 5.5 from the northern part of the Florida peninsula. The diurnal temperature change from 1200 UTC (07:00 local) to 2100 UTC (16:00 local) is shown as the blue contours. The land has warmed by around 10°C over the 9 hours, and the onshore directed flow associated with the sea breeze is evident along both the east and west coasts of Florida. The background synoptic-scale flow is from the southeast and highlighted by the arrow in the lower right of the figure. The observed winds at the two coasts are consistent with the impacts of the background flow on the sea breeze. On the east coast, where the background flow is onshore in the direction of the sea breeze, a stronger onshore flow is observed at the coast. On the west coast, where the background flow is offshore opposite to the sea breeze, weaker onshore sea breeze winds are observed. The approximate location of the sea breeze

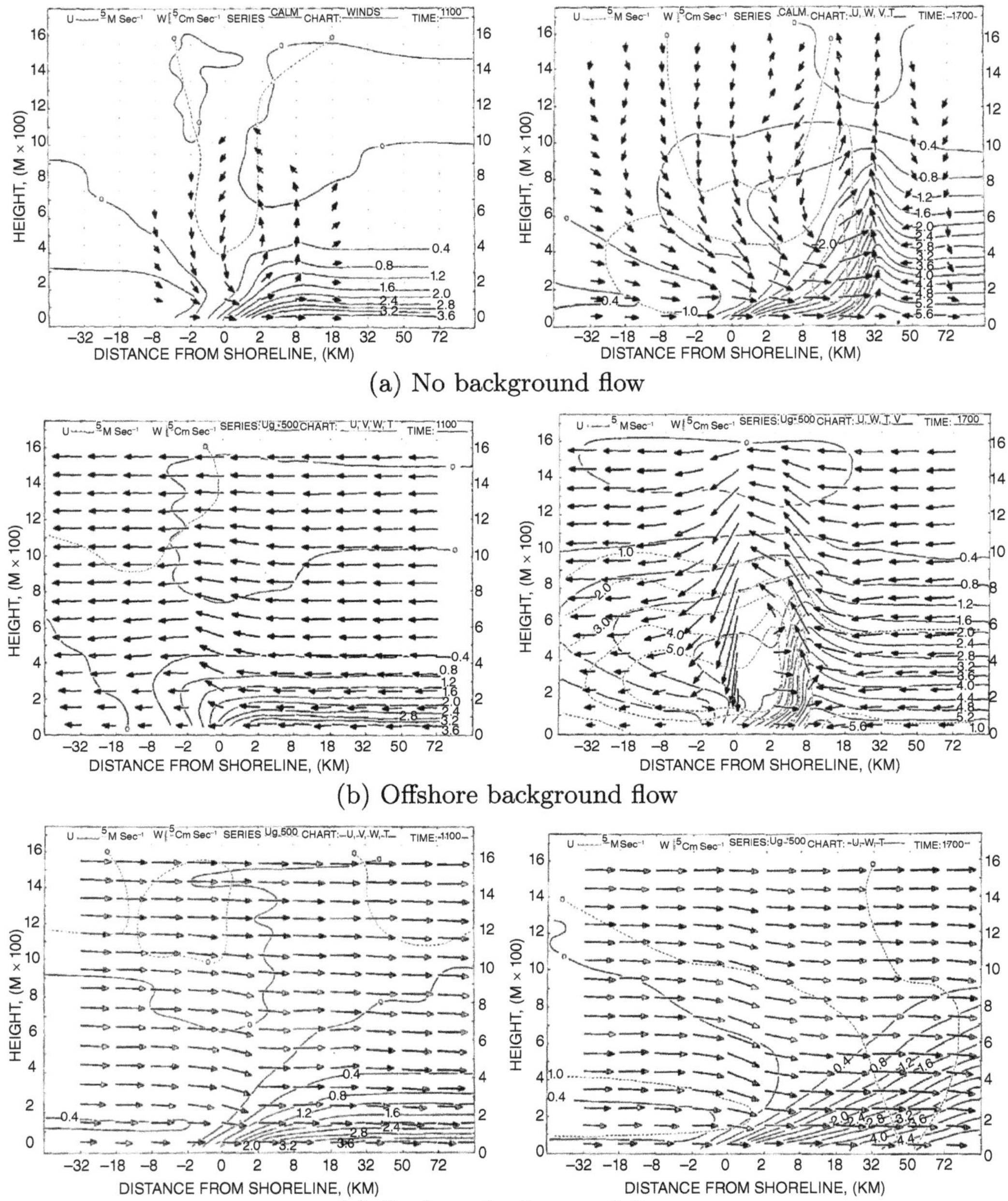

(a) No background flow

(b) Offshore background flow

(c) Onshore background flow

Fig. 5.4 Simple idealized model simulations of sea breeze from Estoque (1962) for differing background flows. Left panels are at 11:00 local time, and right panels are 17:00 local time. (a) basic sea breeze circulation with no background flow, (b) sea breeze circulation with 5 m/s onshore flow, and (c) sea breeze circulation with 5 m/s offshore flow. Published in 1962 by the American Meteorological Society.

fronts for the east coast (heavy dotted line) and west coast (heavy dashed line) are also shown. The primary effect of the background flow is to produce greater inland penetration with the onshore flow (east coast) and less

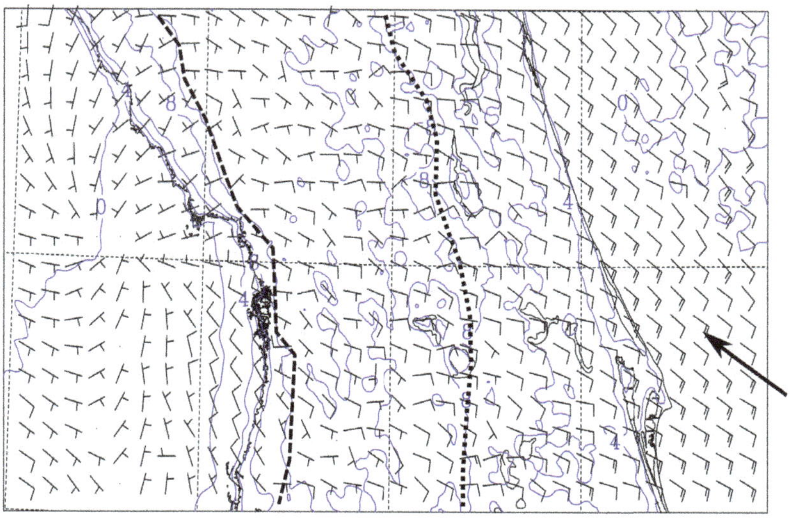

Fig. 5.5 Surface winds (barbs) and diurnal temperature rise (blue contours) in °C for 2100 UTC (16:00 local) for the Northern Florida peninsula. The synoptic-scale background flow direction is indicated by the heavy arrow in the lower right. The approximate location of the sea breeze front for the west coast (heavy dashed line) and east coast (heavy dotted line).

inland penetration with offshore flow (west coast). Given that the coastline does not have the same orientation to the background flow along the entire sections of coast, there are also south-to-north differences. In the north, the background flow is more parallel to both coasts and consequently the west coast front moves inland more than further south, where the background flow is more cross-coast. On the east coast, the coastal orientation to the background flow is more uniform but some north-to-south differences also occur. On the east coast, the impact is seen with the front moving inland more to the south than further north, where the flow is more coast parallel.

5.3.2 Static Stability

The static stability also plays an important modifying role in the sea breeze circulation. The static stability can do any of the following things: damp or enhance vertical motion, enhance or weaken vertical mixing of winds in the boundary layer, and limit the vertical extent of the heating and circulation. The first and last effects are tied together in that, without strong vertical motion, the depth of the circulation tends to be limited. However, the stability structure is important in considering these effects. A strong capping inversion may provide the basic vertical limitation to the surface heating and resultant sea breeze circulation. For example, if an inversion occurs at 200 m, then the depth of the heating is stopped at this level, and the shallow layer will undergo a large temperature rise. The large temperature rise favors a stronger sea breeze even though the vertical motion is damped by the inversion. This is in contrast to a more weakly stratified layer with no inversion, where the heating gets distributed through a deeper layer and the average temperature rise is less. This would result in a weaker sea breeze but stronger vertical motion given the weaker static stability. In addition, more stable stratification below the inversion may weaken vertical motion in the boundary layer even more compared to a weakly stratified boundary layer. The below inversion stratification plays an important role in the vertical distribution of winds in the sea breeze. A well-mixed neutral-to-unstable boundary layer will produce a deep layer of strong sea breeze winds. A stably stratified layer will tend to keep the sea breeze winds strongest aloft. The boundary layer stratification undergoes considerable diurnal fluctuation from very stable at night to unstable in the late afternoon. Hence, it is not uncommon to fail to see a land breeze in surface observations due to the decoupling of the surface layer winds from the rest of the layer.

5.3.3 Vertical Wind Shear

Vertical wind shear of the background flow can substantially alter the character of the sea breeze circulation. Typically, the wind speeds are greater above the surface (due to friction) with little directional shear in the lower atmosphere. This impacts the sea breeze circulation by reducing the offshore flow aloft and strengthening the near-surface flow when the background flow is onshore. If the flow is offshore with stronger flow aloft, the offshore flow in the upper part of the circulation is strengthened, while the near-surface sea breeze (onshore flow) is only weakened slightly due to the weak opposing flow. These are depicted in Fig. 5.6.

Vertical shear may also include directional shear through the lowest layers if warm or cold advection is occurring. In this situation, if there is warm advection (veering) of onshore flow, the sea breeze (lower branch) is strengthened with minimal impact on the return flow aloft, except that it is turned to the right. If the warm advection occurs with offshore flow at low levels, then the sea breeze is weakened, and the return flow remains the same speed but is turned to the left. Under cold advection (backing), if the lower branch is onshore, the sea breeze is strengthened and the return flow speed is unaffected, while the direction is turned to the left. If the lower branch is offshore, the sea breeze is weakened and the upper return flow is turned to the right. The degree of rotation between the sea breeze and return flow depends upon the magnitude of the directional shear or advection in the layer.

5.3.4 Clouds

Clouds can play an important role in the sea breeze as both a modifying influence and an important consequence of the sea breeze. The basic

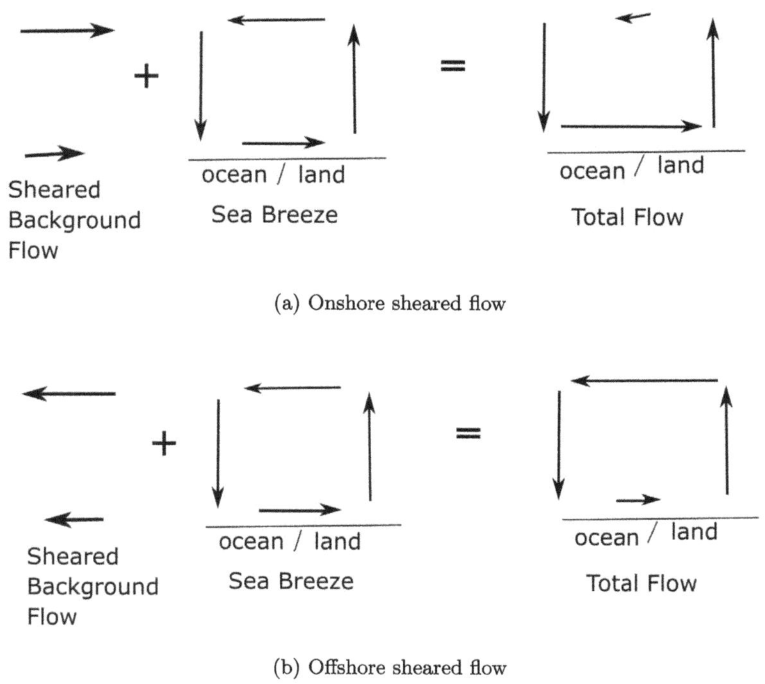

(a) Onshore sheared flow

(b) Offshore sheared flow

Fig. 5.6 Wind Shear influence on Sea Breeze. Vertical sheared background flow is shown on the left for (a) onshore and (b) offshore flows with stronger flow aloft. When linearly added to the sea breeze circulation, the circulation on the right occurs. The low-level part of the circulation is enhanced in (a) as the background adds to surface flow but weakens the upper-level flow. The opposite occurs in (b) where the upper-level is enhanced and the surface flow is weakened.

impact of clouds is to reduce the radiative heating of the land areas during the day and limit their cooling at night. Consequently, clouds tend to damp the heating cycle if they persist through the entire cycle. Strong sea breezes occur on days with sun and strong surface heating. However, the presence of clouds within the boundary layer can play an important role in the timing of the sea breeze. Boundary layer clouds can form from both large-scale flow patterns as well as from the sea breeze circulation itself. Low-level clouds that form inland at night are often the result of cooling the moist air advected inland by the sea breeze on the previous day. This pattern is depicted in Fig. 5.7(a), where the clouds occur overland within the land breeze flow. This boundary layer stratus over the land may initially limit daytime heating, and then once the clouds burn off, the heating may rapidly increase to produce a similar thermal gradient to that with no clouds. The sea breeze onset is delayed in this case but may reach similar intensities to clear days. Boundary layer clouds can form offshore and extend inland as depicted in Fig. 5.7(b). This cloud pattern is most typical along continental west coasts, where a cloud-filled marine boundary layer forms due to the large-scale processes associated with an offshore anticyclone. In this situation, the overland warming may be delayed, similar to the nocturnal low-level clouds noted above. However, depending on the layer depth, the warming can be insufficient to burn off the clouds and result in an overall weaker sea breeze. If clouds do clear over land but not over the water, they may enhance the thermal gradient due to their presence and persistence over the water. This is consistent with very strong sea breezes along the California coast in North America on days when coastal stratus is present that subsequently burns off in the adjacent land areas. The other important relationship between clouds and the sea breeze is the tendency of the sea breeze to produce clouds over the land areas as depicted in Fig. 5.7(c). The vertical motion at the sea breeze front can produce fair-weather cumulus to severe convective clouds due to the focused lifting and the moist potential stability of the inland air. If deep convection occurs, the sea breeze circulation is disrupted from its typical evolution due to the strong vertical accelerations in the convection and associated downdrafts that severely modify the coastal thermal structure.

5.3.5 Coastal Geometry

The coastline characteristics including the basic geometry of the shoreline and the presence of coastal topography also impact the evolution of the classic sea breeze. The coastline geometry plays a particularly important role in the horizontal structure of the sea breeze in the along-coast direction. Given that the thermal gradient must essentially develop in a direction perpendicular to the local coastline (isotherms parallel to the shore), the local sea breeze forcing will always be directed in a perpendicular direction across the coast. Consequently, coastlines that curve in a concave manner toward the ocean such as a bay will tend to produce a sea breeze that is divergent over the land areas, as illustrated in Fig. 5.8(a). Under these circumstances, the upward vertical motion is weakened or eliminated, and cloud formation is suppressed. Coastlines that curve in a convex manner toward the ocean such as a headland or peninsula will tend to produce a sea breeze that is convergent over the land areas as seen in Fig. 5.8(b). Under these circumstances, the upward vertical motion at the sea breeze front will tend to be enhanced and cloud formation is favored. An important aspect of whether the local coastline curvature may be important is the along-coast length scale of the curvature (or radius of curvature). The temperature perturbation of small bays or headlands may simply get mixed out horizontally without producing an identifiable bend in the thermal gradient. In these situations, the effect of the bays or headlands is to produce along-coast changes in thermal gradient intensity without any directional change. The result will be sea breeze intensity changes in the along-shore direction and no impact on the sea breeze convergence due to directional differences. Under these circumstances, the strongest sea breeze will produce the strongest divergence or convergence and will be the preferred centers of clearing or cloud development, respectively.

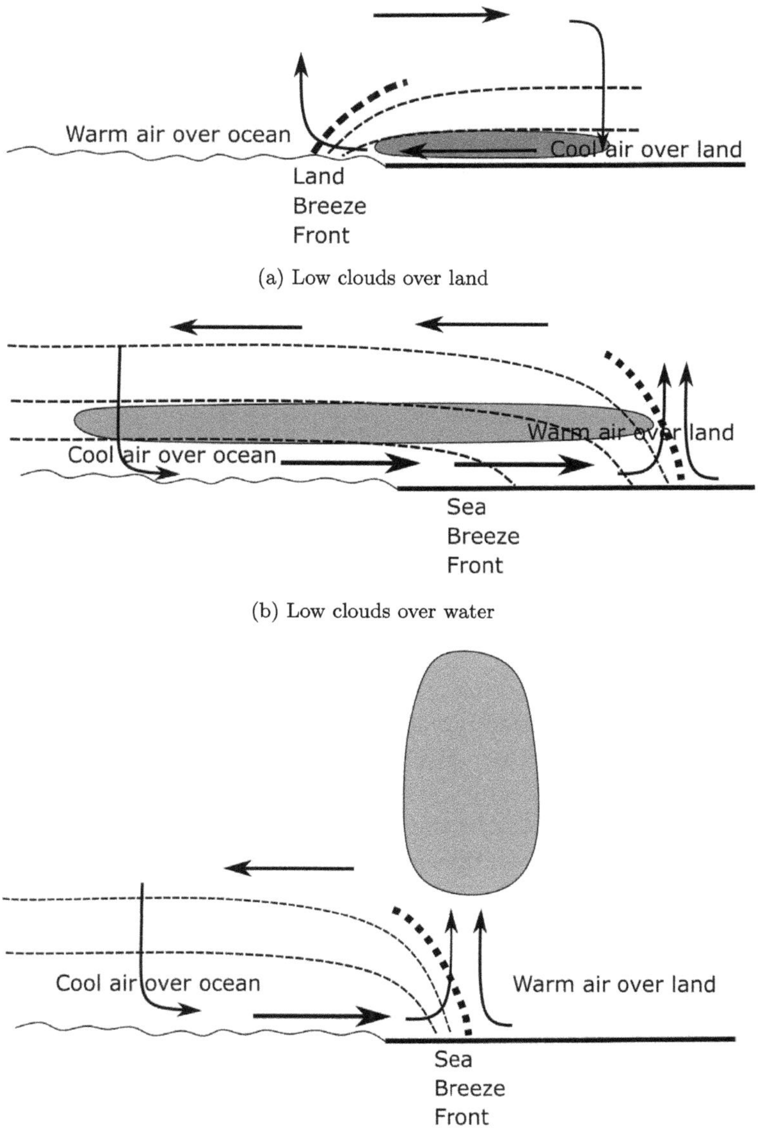

(a) Low clouds over land

(b) Low clouds over water

(c) Clouds forced by sea breeze

Fig. 5.7 Cloud influence on Sea Breeze. Arrows represent sea breeze circulation, thin dashed lines are isentropes, and shaded areas represent clouds. Potential cloud patterns for (a) low-level clouds inland at night, (b) low-level clouds over the water and advected inland, and (c) clouds forced by rising motion at sea breeze front.

5.3.6 Topography

The other coastline characteristic that impacts the sea breeze is the presence of coastal mountains. Topography can have several important effects on the sea breeze. The first effect is that the sloping coastal terrain has a

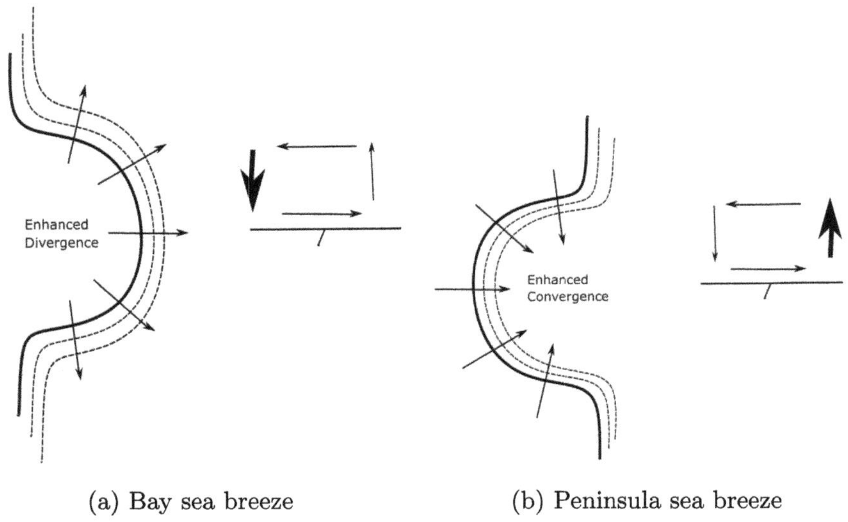

Fig. 5.8 Influence of coastal geometry on sea breeze. Coastline is solid line, dashed lines are isotherms, and arrows are cross-coast sea breeze flow at the surface. Vertical cross-section of idealized sea breeze circulation is shown for (a) bay or (b) peninsula circulations. Thick arrow represents stronger vertical motion.

tendency to heat up more rapidly than a flat coastline. This more rapid heating is due to the greater effective surface area over which, early morning sun strikes. This effect is the well-known mountain-valley circulation, which is a land example of a thermally driven circulation. The impact of this additional heating on the coastal sea breeze is to produce a stronger sea breeze earlier in the day than might otherwise be expected. The absolute intensity might not be that different since the amount of heating over the heating period is not that different, just the time scale over which it occurs. Hence, mountainous coastlines will have a preference for early sea breeze onset compared to the adjacent flat valleys. In addition, the topography can also prevent inland movement of the sea breeze front if the slope is sufficiently steep. The other impact of the topography is its ability to shape the horizontal flow through channeling effects. The sea breeze is being forced in a direction perpendicular to the coastline. A valley that lies at some angle relative to the coastline will have a tendency to channel the sea breeze down or up the valley. Consequently, the sea breeze may start in a nearly cross-coast direction but then be rotated into an along-valley direction as the length scale of the sea breeze increases. This situation is depicted in Fig. 4.9 where the thermally driven circulation aligns with the Salinas Valley. The valley is oriented in a direction that follows the large-scale coastal orientation. The initial sea breeze in the Monterey Bay is from the west but then rotates to the northwest to align with the topographic valley due to channeling by the topography. The general problem of topographic influences on the flow will be considered in more detail in Chapters 6–11.

5.4 Exercises

5.1 The simplified sea breeze circulation model described in the text suggests that the wind speed along the lower branch of the circulation is essentially uniform. However, observations indicate that the maximum surface wind speeds actually occur in a narrow zone along this lower branch at the sea breeze front. On a vertical cross-section, draw the typical temperature and wind structure of a sea breeze circulation. Identify the zone of maximum low-level winds and write down the physical basis for this occurrence.

5.2 Limited ocean channels (short distance across) and lakes tend to produce less intense sea breezes than open coastlines with similar water temperatures. Draw a diagram of the circulation in this situation and discuss why this circulation is more self-limiting. Hint: Consider the influence of the downward branches of the circulation.

5.3 The sea breeze is driven by the differential heating of land versus water through a layer whose depth depends on the vertical structure and amount of heating.

- Calculate the depth and average temperature rise through which heating will penetrate, assuming a constant background stratification of 10 K/km and a surface temperature rise of 7°C.
- Assuming that the above-stated temperature rise is linear with respect to time, calculate the average gravity current phase speed (from initiation to max heating) and determine the inland penetration that will occur for this sea breeze front.
- Given the inland penetration and depth calculated above, use the formula presented in the text to determine the maximum sea breeze wind speed. Assume inland penetration is only half the length scale of the full sea breeze circulation.

5.4 Synoptic-scale background flow can greatly alter the sea breeze structure and evolution. Weak offshore flow (less than 8 $\mathrm{ms^{-1}}$) tends to strengthen the sea breeze front but also limit its inland propagation. Using an offshore flow of 5 $\mathrm{ms^{-1}}$ and the gravity current phase speed ($c = (h\frac{g\delta\theta}{2\theta})^{\frac{1}{2}}$) determine how large the thermal gradient must be to get a 2 $\mathrm{ms^{-1}}$ speed of advance of the sea breeze front. Assume the depth of sea breeze inflow is 500 m.

5.5 While the forcing of the land breeze is qualitatively similar to the sea breeze, the structure of the boundary layer over the land evolves differently when the surface is cooling versus warming. Sketch a representative cross-coast cross-section of potential temperature and winds for the sea breeze and land breeze. Explain any differences between the two in terms of stability and depth of the thermal perturbation. Comment on whether the sea breeze or land breeze winds are stronger and what leads to these differences.

5.6 The sea breeze evolution described in the text assumes that we can neglect the impact of the Coriolis force through the life cycle of the sea breeze. For a persistent sea breeze that lasts closer to an inertial cycle ($\frac{2\pi}{f}$), then the wind responds to the Coriolis force and turns more perpendicular to the pressure gradient. Discuss which direction the winds will turn along the east and west coasts of Florida due to the Coriolis force. What impact would this rotation of the wind have on the propagation of the sea breeze front, and why?

5.7 Explain the following statements:

- A stronger sea breeze occurs under weakly stratified conditions compared to strong static stability.
- Surface heating is not always a good indicator of sea breeze strength.
- Deep inland penetration of the sea breeze occurs when there is a deep layer of very cool marine air.

5.8 The cross-coast perturbation pressure gradient caused by preferential heating over the land is assumed to drive the wind field.

- Calculate the cross-coast pressure gradient you would get for a 2°C average heating through 1000 m (assume the top of the layer is 875 hPa).
- Since we generally observe minimal surface pressure gradient across the coast, explain how the offshore-directed return flow acts to weaken or eliminate the surface pressure gradient. You might sketch a sea breeze circulation and consider what the pressure gradient aloft must do.
- Given that the return flow aloft can be distributed through the vertical, how would this effect your reasoning given in the previous part.

5.9 Consider an island 50 km across where the land heats at a rate of 2°C/h from sunrise until the sea breeze front passes. Estimate the time before sea breeze fronts from opposite coasts will meet and cause the sea breeze to decay. Assume the depth of the sea breeze is 300 m and that the time varying potential temperature difference across the sea breeze front is given by the heating rate.

5.10 Derive how much the vertical motion offshore will change if the sea breeze occurs in a 10 km wide bay versus a straight coastline. Assume the sea breeze is 5 ms^{-1} and derive the vertical motion through the continuity equation, assuming the sea breeze layer is 1000 m deep (downward motion starts at 1000 m).

Low-Level Coastal Jets 6

Coastal jets refer to the occurrence of strong low-level winds that can occur along coastlines. Typically, the wind maximizes just above the surface and blows in an along-coast direction. These strong winds can impact the underlying ocean to produce coastal upwelling as observed along the west coasts of North and South America. In addition, the strong flow along the coast can impact the coastal vegetation by limiting inland transport of moisture as well as creating a persistent wind-swept environment in the immediate coastal region.

6.1 Description of Coastal Jets and Types

Coastal low-level jets and indeed jets in general occur due to a persistent baroclinic structure. Along coastlines, this baroclinic structure tends to arise due to the differential surface heating that occurs across the coast. While this heating gradient typically drives the thermally driven sea breeze circulation, if the gradient can persist longer than the diurnal cycle, then an along-shore flow develops into a jet like structure. To maintain a cross coast thermal gradient, either the large-scale flow must be essentially coast parallel or coastal topography must disrupt the inland advection of marine air.

One of the most notable coastal jets is the jet that occurs along the US West Coast during the summer months. The US West Coast during the summer months is dominated by large-scale subsidence associated with the climatological Pacific high-pressure center. The climatological distribution of sea-level pressure produces persistent northwesterly surface winds over the eastern Pacific Ocean. These persistent winds result in forcing strong coastal upwelling to produce very cool ocean temperatures near the coast. Consequently, a strong low-level inversion and a well-developed marine boundary layer (MBL) are established over the region.

6.2 Large-Scale Structure and Geostrophic Considerations

To begin to understand the coastal jet, it is first useful to describe what the jet actually is and what its structure typically looks like. If we consider

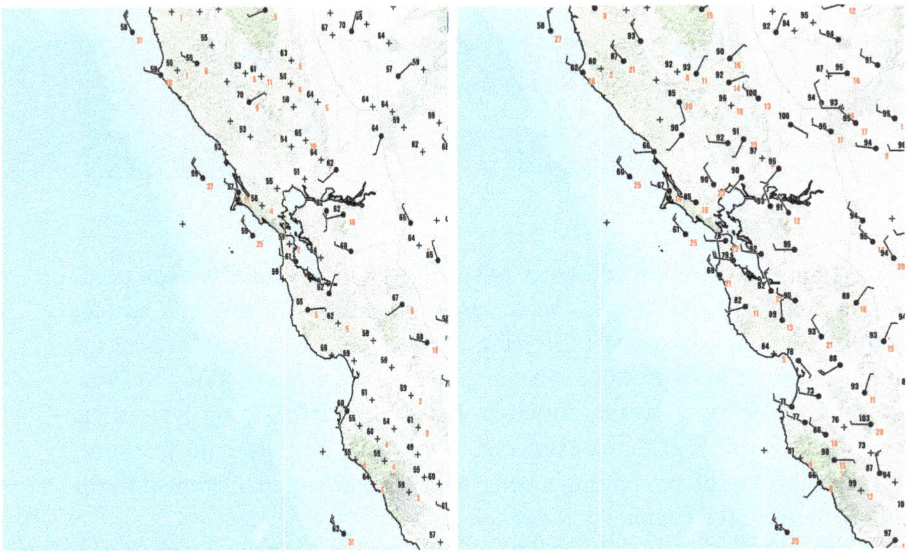

(a) Surface observations at 1200 UTC (b) Surface observations at 0000 UTC

Fig. 6.1 Surface observations at (a) 1200 UTC (early morning) and (b) 0000 UTC (late afternoon) for the California coastal region. The observations show evidence of the coastal jet with strong winds offshore compared to over land. The strongest winds occur along the coast at both times.

the observed surface winds at 1200 UTC and 0000 UTC as depicted in (Fig. 6.1) for the California coastal region of North America, it is evident that the wind speeds are generally higher offshore than over land and that this difference persists throughout the day. On this particular day, the land stations show 5–10 knot winds, while the coastal buoys have winds of 15–25 knots depending upon where along the California coast we look. There is little difference between the 1200 UTC and 0000 UTC winds at the buoy locations, suggesting that any diurnal variation is small. Without a doubt, one difference between the surface winds over land versus those over the water is the difference in friction between land and water. Although this friction difference is important in determining the surface wind speed, the more fundamental difference in the winds is found in the pressure field (Fig. 6.2), which shows a pronounced cross-coast pressure gradient near the coast in this Climate Forecast System Reanalysis (CFSR) analysis for a particular day in July 2014. This pressure gradient is a maximum at the coast (even at coarse resolution), and it typically decreases both landward and seaward of the coastal boundary. Also depicted in Fig. 6.2 is the near-surface wind isotachs at 950 hPa. The fastest winds are coincident with the strongest pressure gradient, indicating a highly geostrophic response. The dynamics will be considered more thoroughly in Section 6.4.

The factors that contribute to the strong cross-coast pressure gradient are most evident in a cross-section of the structure across the coast. A cross-section of potential temperature for a day in July,

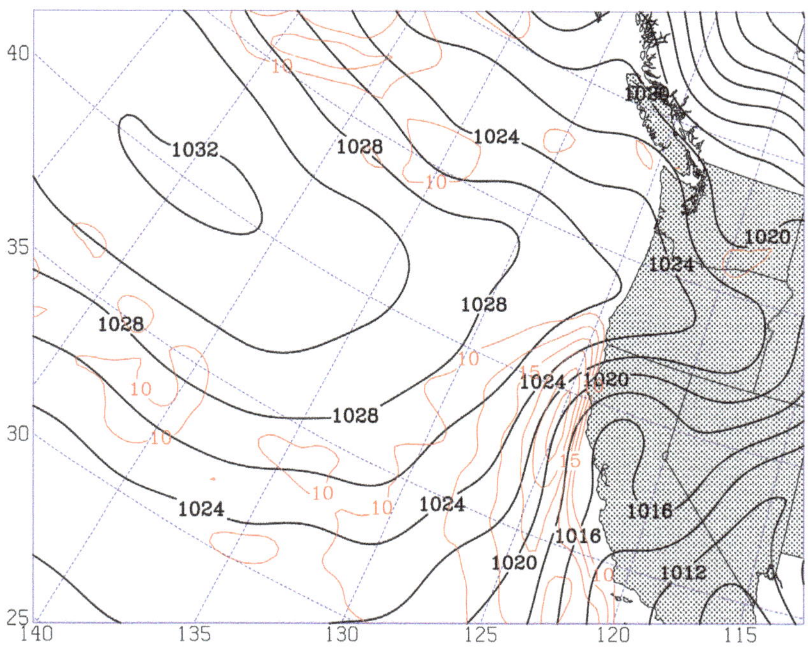

Fig. 6.2

Surface analysis from CFSR global model showing near-surface wind isotachs (red) and sea-level pressure (black) for 1800 UTC July 24, 2014.

shown in Fig. 6.3, shows the potential temperature lines sloping down at the coast, indicating strong baroclinic structure and a relatively well-mixed MBL offshore with constant potential temperature in the vertical. The depth of the MBL increases seaward, which is typically observed (Neiburger, 1961). This increase in depth is due to a combination of weakening large-scale subsidence further west and generally warmer sea surface temperatures (larger surface heat flux) well offshore. The sloping isentropes due to the deepening MBL offshore result in the observed baroclinic structure, which through thermal wind considerations (or hydrostatic considerations) leads to a low-level wind maximum and the strongest cross-coast pressure gradient at the surface. This yields the jet structure seen in Fig. 6.3, which is the so-called California coastal jet. The wind maximizes above the ground due to the effect of friction retarding the wind closer to the surface. A schematic is shown in Fig. 6.4 that shows the inversion sloping down toward the coast and the surface winds increasing toward the coast which is typically observed.

The coastal jet depicted in Fig. 6.3 is associated with the larger-scale baroclinic structure of the atmosphere over the California region during the summer. This is evident by considering the climatological 500 hPa, 850 hPa and surface analyses, shown in Fig. 6.5. At 500 hPa, a weak trough occurs over the west coast and produces southwesterly flow aloft (about 10 m/s). The thermally induced surface low over the desert Southwest

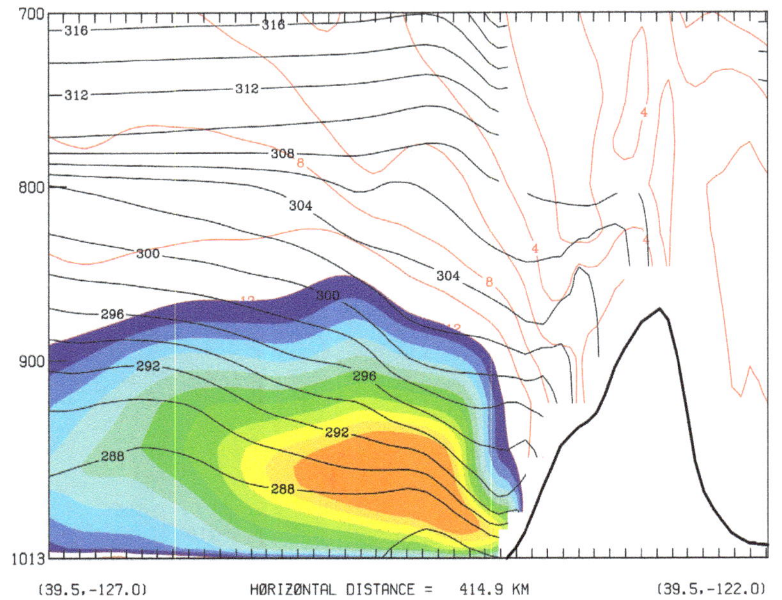

Fig. 6.3 Cross-section of thermal and isotach structure across the coast in a model simulation of a coastal jet for 0000 UTC July 21, 2021. Black contours are potential temperature plotted every 2K. Red contours and color fill are isotachs of the flow perpendicular to the cross-section plotted every 2 m/s. Maximum winds are approximately 24 m/s.

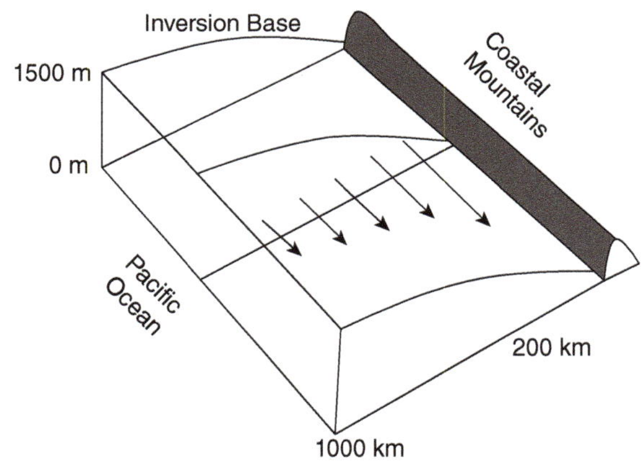

Fig. 6.4 Schematic showing structure of sloping inversion and low-level winds associated with coastal jet.

and interior part of California result in a pronounced west-to-east pressure gradient force and the persistent northwesterly winds at the surface (about 8–10 m/s). This implies a thermal wind vector that points to the north-northeast up the coast over a broad region by comparing the surface wind in Fig. 6.5(c) to the upper-level winds in Fig. 6.5(a). The weak

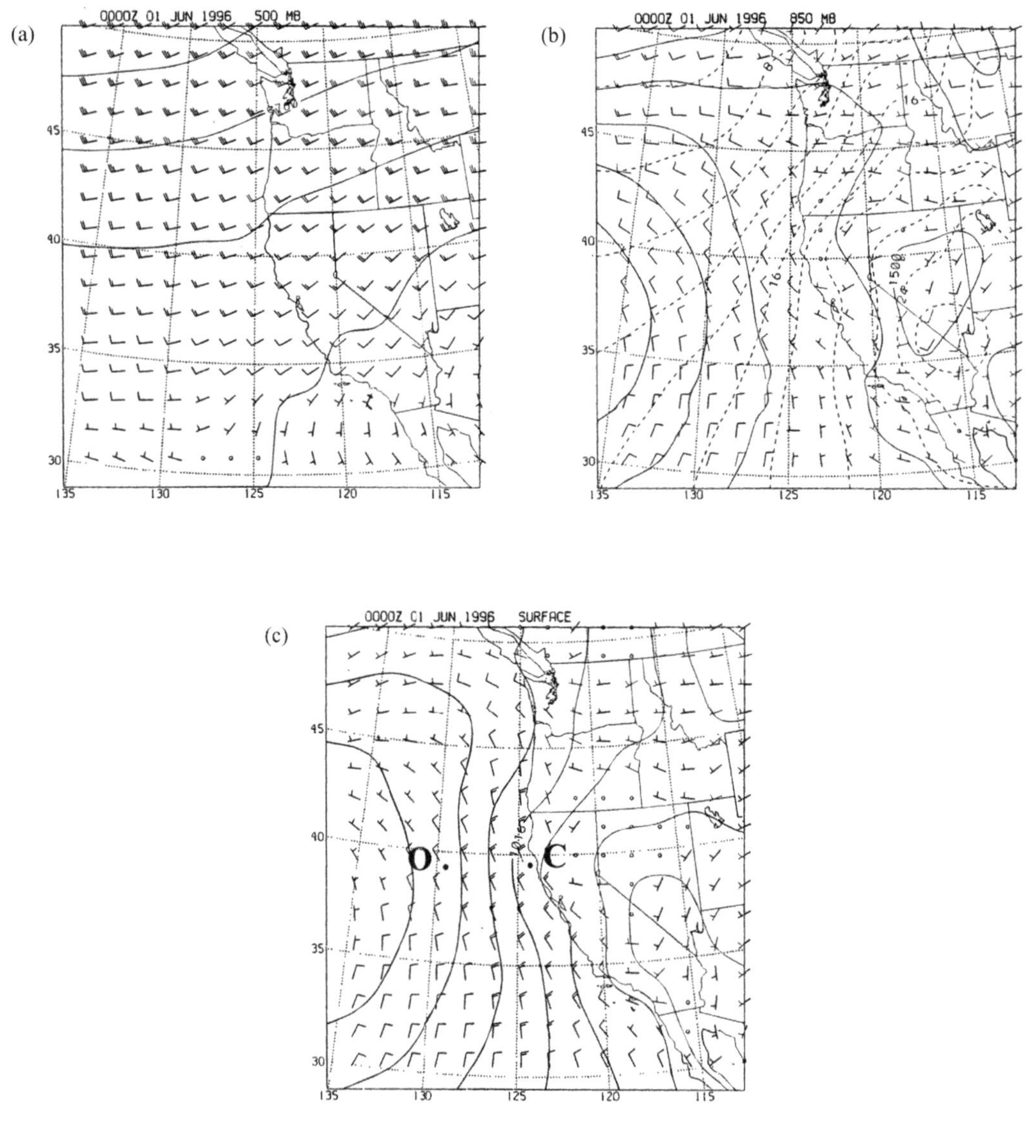

Fig. 6.5 Synoptic composites of the structure associated with coastal jet conditions at a) 500 hPa, b) 850 hPa, and c) the surface. The locations marked O and C indicate locations offshore and along the coast where the vertical shear in the lower atmosphere is compared.

height gradient over the coast at 850 hPa yields weak northwesterly flow at this level that must increase downward to produce the strong northwesterly flow at the surface. This implies a thermal wind vector in the surface to 850 hPa layer that points up the coast opposite the surface flow in a northwesterly direction. The thermal structure in the 500–850 hPa layer produces a thermal wind vector from the southwest (similar to the 500 hPa winds) that increase the flow from the southwest. This jet is a broad feature that extends 300–500 km offshore as seen in the surface wind plot in

Fig. 6.5(c), where 15–20 kt winds are found to extend offshore to about 500 km. Detailed observations near the coast taken with aircraft show that the jet core can be considerably more compact than that implied by the large-scale structure. The jet can be only 20–40 km wide in some places, which makes it a true coastal jet. The mechanisms that produce this mesoscale jet depend on aspects of the coast not embodied in the broad baroclinic structure of the large-scale climatological structure.

If we consider the thermal wind structure more carefully, it is evident that the coastal jet at low levels is primarily due to the cross-coast thermal gradient in the lowest levels. If the thermal wind in the 500–850 hPa layer is considered in Fig. 6.5, then it is evident that a north-northeasterly directed thermal wind vector occurs, similar to that found for the surface to 500 hPa layer. This is associated with the colder temperatures over the ocean to the northwest of California, which produces the 850 hPa and surface high offshore below the weak 500 hPa trough. The 850 hPa structure is very similar to that at the surface with some slight differences in the strength and placement of the synoptic-scale features. Now consider the thermal wind or vertical wind shear between the surface and 850 hPa. Further offshore at point O in Fig. 6.5, say near 130 W, the 850 hPa flow is about 10–15 kts from the northwest, and the surface flow is about 10 kts from the northwest. This implies a thermal wind near zero or perhaps from the northwest and not very large, which suggests a rather uniform thermal structure between the two layers. Near the coast at point C in Fig. 6.5, the 850 hPa flow is weak and still from the northwest, while the surface winds are strong from the northwest. This implies that the thermal wind vector is from the southeast (up coast) and fairly large, which implies that there is a strong cross-coast thermal gradient with warm air inland. The strong cross-coast thermal gradient is primarily confined to the surface to 850 hPa layer and represents the residual sea breeze over California as the cross-coast pressure gradient does not decay at night due to the coastal mountains that prevent the inland penetration of cool air. The northwesterly surface flow is the geostrophic response to this sea breeze (thermally produced) pressure gradient, which persists for periods longer than the diurnal cycle. The cross-coast structure depicted in Fig. 6.6 looks like a sea breeze type structure with warm air over the land and cooler air over the water that produces an onshore surface pressure gradient. The thermal structure increases the thickness over the land and reduces it over the water to produce a weak pressure gradient at 850 hPa. The coastal jet owes its basic existence to the persistence of this structure. The detailed structure of the marine inversion becomes a key element in defining the jet structure as the slope of the strong inversion dominates in determining the thickness between the layers. Putting the two layers (850 hPa-surface and 850 – 500 hPa) back together, it is evident that there is a thermal gradient

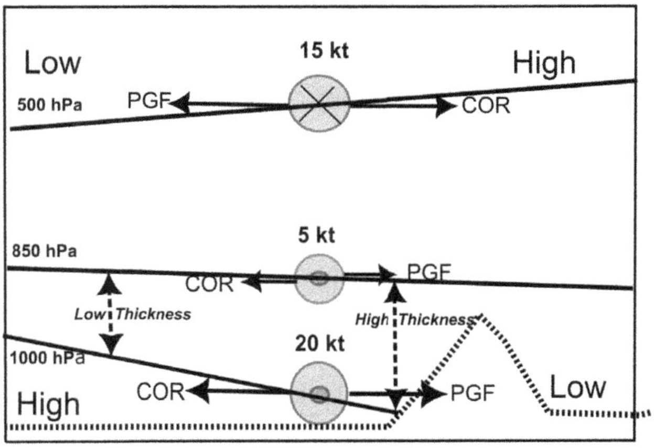

Fig. 6.6

Cross-section of idealized thermal structure across the coast in a coastal jet. The thickness in the 1000 — 850 hPa layer increases over the land due to warming. This results in an increasing cross-coast pressure gradient as you move down to the surface. The wind speeds increase toward the surface due to the changing pressure gradient, which represents thermal wind balance.

aloft that turns the wind down coast, but it is really the low-level thermal gradient that makes this into a jet structure and this low-level thermal structure is definitely a coastal effect.

The thermal structure definitely plays the dominant role in determining the presence of the coastal jet; however, topographic effects may also contribute to its intensity. Consider the low-level jet, which typically has a small onshore flow component due to the frictional turning toward low pressure. The presence of the strong low-level inversion tends to block this component of the flow and will turn this component down the pressure gradient, which is typically down the coast. The flow will accelerate most strongly near the coast and decay out toward the Rossby radius of the mountains, which leads to stronger winds near shore and weaker offshore consistent with the observations and the structure implied by the thermal gradient as well. The key aspect of this interaction is that the flow must be parallel to the coastal mountains, and its strength will be partially determined by a balance between friction and the down-coast pressure gradient (unless the winds are unbalanced and accelerate down coast). Results from Cross (2003) indicate that while thermal wind balance is dominant in the coastal jet, the strongest jets tend to occur when the synoptic scale pressure gradient is rotated slightly up-coast to allow for down-coast acceleration of a small blocked component. The relative importance of these effects (topographic blocking and thermal wind balance) has not been fully documented in the literature and probably varies from day to day.

6.3 Mesoscale Along-Coast Variations

An important aspect of the California coastal jet is its along-coast variablity, which impacts the distribution of clouds, winds, and its ocean forcing. This along-coast variation is evident in Fig. 6.7, which depicts the low-level isotachs from model simulations of a case from July 2021. The isotach maxima generally occur downwind of coastal points and capes, where the orientation of the coastline and coastal mountains changes substantially. From the perspective of the basic thermodynamic forcing of the jet, these regions of stronger flow should have a stronger cross-coast thermal gradient. However, these regions along to coast do not necessarily have any stronger thermal gradients than elsewhere along the coast. Although variations in the cross-coast thermal gradient occur along the coast, they do not correlate with the strongest winds, indicating that other processes than thermal wind balance are occurring.

The development and evolution of these along-coast maxima have been documented observationally, and a number of dynamic explanations have been offered to explain their existence. The most common explanation is based on the adjustment of super-critical flow to bends in the coastal mountain barrier (Winant et al., 1988), which will be described in more detail in Section 6.4. However, mountain waves and thermodynamic explanations have also been considered. Detailed observations of the jet

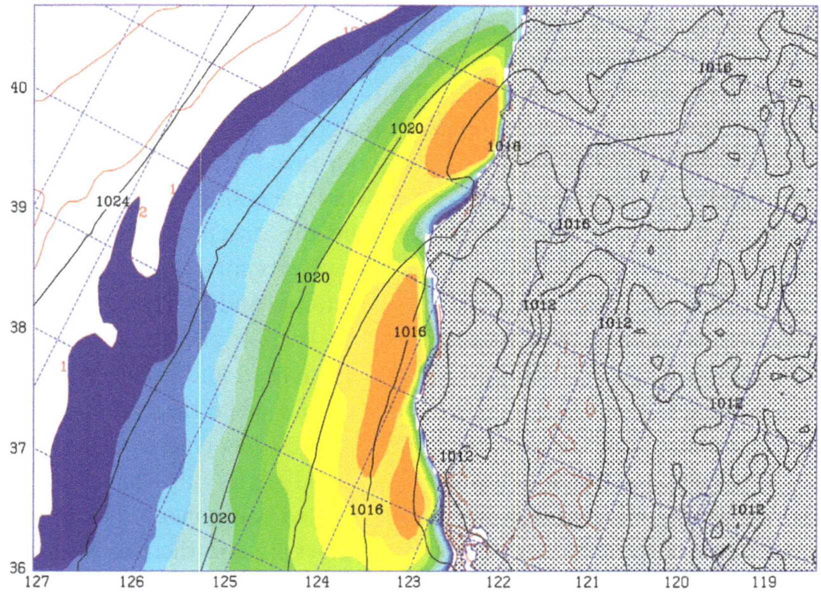

Fig. 6.7 Horizontal structure along the coast in a model simulation of a coastal jet from 0000 UTC July 21, 2021. Wind speed maxima occur downstream (south) of points and capes along the coast.

structure near Cape Mendocino and Pt. Sur indicate that the inversion is typically low (well below the coastal mountains) and has a strong slope down toward the coast. The steep inversion slope and reduced boundary layer depth do not uniquely determine the possible dynamic explanations resulting in some ambiguity about why the wind maxima occur and how they may evolve.

6.4 Supercritical Effects

To begin to explain how wind maxima occur near points and capes, it is useful to consider the detailed structure of the flow in these regions under several situations. Winant et al. (1988) consider three basic synoptic settings and compare the structure of the MBL in each of these cases. The three patterns are shown in Fig. 6.8 and can be described as (1) weak synoptic scale flow, (2) post-frontal northwesterly flow, and (3) well-developed inland thermal trough. The first pattern results in weak winds with no particular along-shore variation. The second and third patterns result in relatively strong northwesterly coastal winds but differing mesoscale coastal responses. Winant et al. (1988) examined the MBL structure for these three basic flow patterns. The weak flow (pattern 1) case is characterized by a deep well-mixed MBL up to about 600 m. The difference between patterns 2 and 3 becomes most evident in the low-level thermal structure. Pattern 2 shows a shallow well-mixed MBL and weak stable stratification above, quite characteristic of the post-frontal atmosphere with developing large-scale subsidence. Pattern 3 is the characteristic west coast summer-time MBL, with a well-mixed layer capped by a strong inversion (order of 12K in this case). Both patterns 2 and 3 show a wind maxima at the top of the well-mixed layer characteristic of strong downward momentum mixing and frictional dissipation at the surface.

The detailed structure near Pt. Arena of the low-level flow and sea-level pressure distributions for the three patterns are shown in Fig. 6.9. The weak flow situation (pattern 1) shows weak flow and a very flat pressure distribution as one would expect. Of more interest are the distributions for patterns 2 and 3 which show a relative wind maxima to the south of Pt. Arena. Pattern 2 has a broad maxima and pattern 3 has a considerable small-scale variability with wind speeds ranging from less than 10 m/s to over 20 m/s. The pressure distribution is most instructive and shows that the wind maxima in pattern 2 occurs in the region of the largest pressure gradient, while the pattern 3 maxima and minima occur essentially coincident with the lowest and highest pressures, respectively. The dynamics of pattern 3 was addressed by Winant et al. (1988) and will be subsequently examined. However, it is useful to consider pattern 2 a bit more first. The

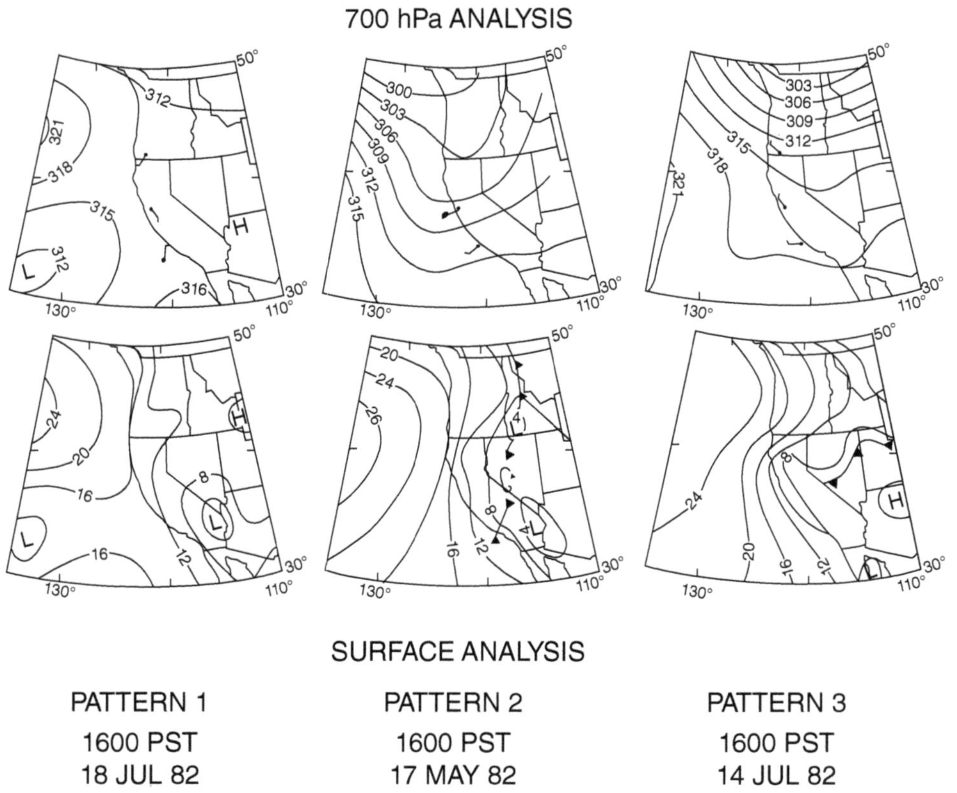

Fig. 6.8 Synoptic patterns of three different days of characteristic coastal flows during the warm season. Pattern 1 represents weak flow aloft at 700 hPa and near the surface in the sea-level pressure pattern. Pattern 2 is associated with an upper-level trough passage and weak surface cold front passage as seen in the sea-level pressure analysis. Pattern 3 has ridging aloft at 700 hPa and a thermal trough extending north through central California. From Winant et al. (1988). ©American Meteorological Society. Used with permission.

higher winds are coincident with the larger down-coast pressure gradient as noted above. This is consistent with the idea that the near-shore wind essentially represents a balance between the pressure gradient and friction, which is characteristic of a blocked cross-coast wind component. Winant et al. (1988) do not calculate the Froude number for this case, but it is probably sufficient for blocking, given the small magnitude of the cross-shore wind component. A more thorough analysis is needed to confirm this hypothesis, but certainly Fig. 6.9 is suggestive. The mechanism that produces this mesoscale pressure gradient was equally unexplained by Winant et al. (1988), but thermal profiles up and downstream of Pt. Arena show a warmer profile downstream than upstream. This results in lower pressure to the south and the along-coast pressure gradient.

The pattern 3 situation is characterized by a strong low-level inversion that is below the tops of the coastal topography in this region. For this quasi-two-layer structure, changes in the MBL inversion base height

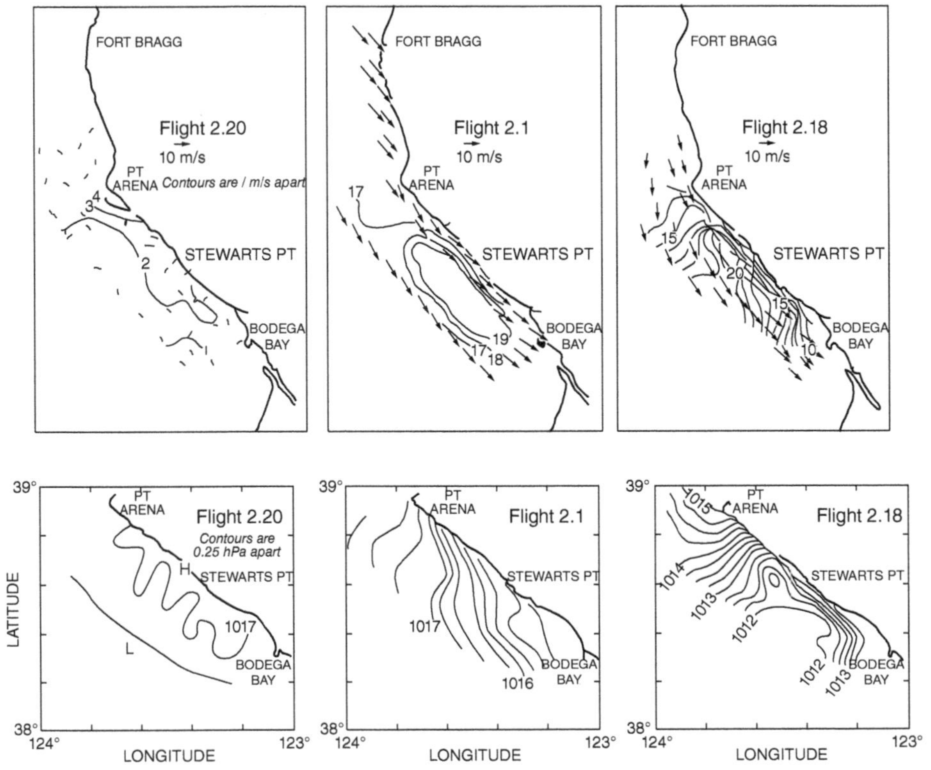

Fig. 6.9 Detailed wind and pressure patterns associated with the characteristic synoptic scale coastal flows. Pattern 1 has weak flow along the coast near Pt. Arena. Patterns 2 and 3 show wind maxima south of Pt. Arena but exhibit different pressure patterns associated with the wind maxima. From Winant et al. (1988). ©American Meteorological Society. Used with permission.

of a 400 m can easily account for sea-level pressure changes of 2.0 hPa which is comparable to those seen in Fig. 6.9. This can be seen by considering the perturbation vertical motion equation 1.5.8c and assuming hydrostatic balance to get $-\frac{1}{\rho}\frac{\partial p'}{\partial z} = g\frac{\theta'}{\theta_o}$, where θ' represents the potential temperature change between the lower and upper layers. This gives the pressure difference as $\delta p' = \rho g \frac{\theta'}{\theta_o}\delta z$. If a layer height change of 400 m, a density of 1.3 kgm^{-3}, and a potential temperature difference of 12°C are assumed, then a pressure difference of 2.04 hPa occurs. If we assume that the MBL height change is responsible for the pressure variation over the region, then the low-pressure regions are associated with low-inversion base heights and the high pressure with high-inversion base heights. Given this structure, then high winds are directly associated with low inversion base heights and vice versa for low winds. This type of behavior is similar to that of hydraulic flow through a pipe, where the flow speed increases when the pipe diameter decreases. The primary difference is that the three-dimensional atmosphere is not necessarily bounded on four sides. For the present situation, this may be true given, mountains to the left of the flow, a strong inversion above, the ocean below, and a large-scale pressure gradient directed onshore to prevent flow away from the coastal channel. Based on these considerations, Winant et al. (1988) make the analogy with hydraulic flow through a pipe. Given this analogy, we can then consider the implications of the coastal bends on the flow along the coast.

To mathematically describe the flow along the coast under conditions that it acts like hydraulic flow in a channel, it is necessary to consider the behavior of gravity waves from a shallow water perspective. The behavior of the flow is governed by a Froude number defined as

$$Fr = \frac{V}{(g'h)^{1/2}}, \tag{6.4.1}$$

where V is the velocity of the flow and $(g'h)^{1/2}$ is the phase speed of gravity waves from the shallow water equations. $g' = g\frac{\Delta\theta}{\theta}$ is a reduced gravity which is defined by the depth of the more dense lower layer h and the density difference across the interface $\Delta\theta$. This phase speed also represents the potential energy required to change the height of the layer interface. If the Froude number is less than 1, then the gravity wave phase speed is greater than the ambient wind speed, and so waves can propagate both up and downstream from their point of initiation. The flow is said to be subcritical in this case. If the Froude number is greater than 1, then the wind speed is greater than the gravity wave phase speed, and the waves travel downstream only since they are being advected downstream faster than they can travel upstream. The flow is said to be supercritical in this case. What initiates gravity waves in the flow? Anything that perturbs the flow will result in a gravity wave response. Certainly, the abrupt change in flow direction that occurs at a point or cape will excite gravity waves at that location, which will propagate out in all directions from the point depending upon whether the flow is sub or supercritical.

Let's consider the dynamics of this adjustment to a change in coastal orientation when the flow is supercritical and the response must occur downstream from the point or cape. Figure 6.10 diagrams the geometry of the situation, where θ is the angle between the upstream and downstream coastal orientations. Positive θ would imply the coastline is turned to the right from its prior orientation and negative would be a turn toward the left as depicted in the figure. As the flow goes past the point, it must turn toward the wall to fill the void, and it excites gravity waves to adjust to this perturbation in the flow. The line extending from the point out to the right of the flow represents the line of furthest upstream propagation of gravity waves as depicted by the expanding wave rings. This occurs when the flow is supercritical. The angle β that this line makes with the coastline is defined simply by

$$sin\beta = \frac{C}{V} = \frac{1}{Fr}. \tag{6.4.2}$$

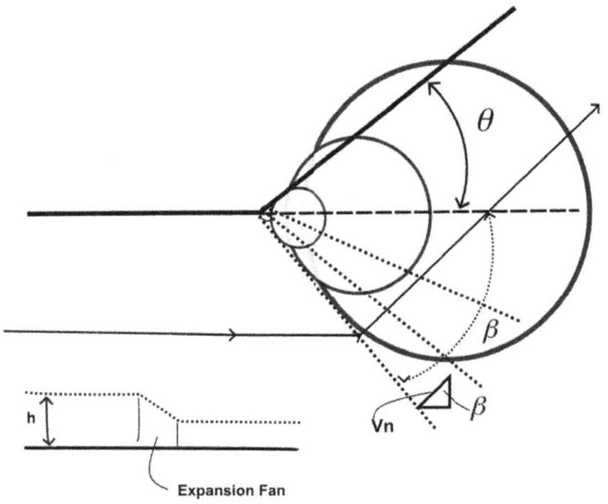

(a) Coastline turns away from the flow

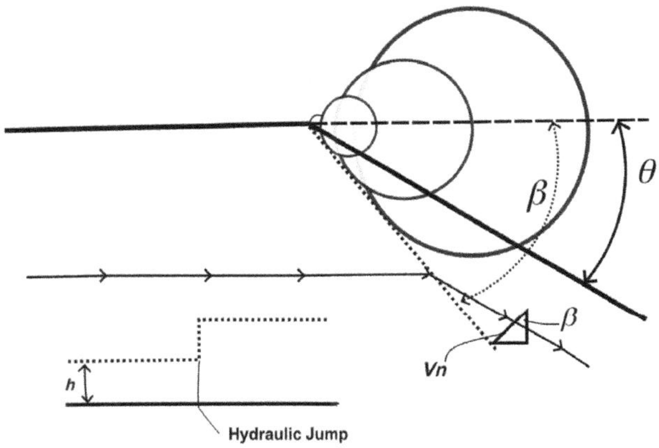

(b) Coastline turns into the flow

Fig. 6.10 Characteristic structure of flow hydraulically adjusting to changes in coastal orientation. Thick solid line represents the coastline which bends (a) away from the flow or (b) toward the flow at an angle θ. The dotted lines represent the upstream edge of the gravity wave propagation or wavefront. The gravity waves are shown by the expanding circles as they get advected downstream from the coastal bend. The angle β represents the maximum upstream wave front.

For the wind to change directions from the point, it must accelerate toward the wall (in the x direction). This acceleration must be $-Vd\theta$, which represents the differential vector that points perpendicular to the original flow direction. If we partition the wind into a component normal to the wavefront (V_n) and parallel to the wavefront (V_p), then it is possible to examine the balance of forces acting on the flow at the wavefront. First, note that the wind parallel to the wavefront is not impacted by the gravity waves and is unchanged. The normal component V_n must change by $dV_n cos\beta$, which is the projection of the normal component into the direction that the wind must turn $-Vd\theta$. The two accelerations must balance which leads to the following expression:

$$dV_n cos\beta = -Vd\theta. \tag{6.4.3}$$

This expression comes from the geometry of the problem and the recognition that the flow across the wavefront must account for the turning of the wind to adjust to the new coastal orientation.

Now examining the momentum equation without Coriolis or friction for the shallow water flow, it is possible to derive an alternative representation of the acceleration across the wavefront. The momentum equation is

$$\frac{dV}{dt} = g'\frac{\partial h}{\partial y} \tag{6.4.4}$$

which defines the acceleration in terms of the pressure gradient in the y-direction that must be given by changes in h for the shallow water equations. The momentum equation implies that

$$g'dh = VdV, \tag{6.4.5}$$

which when it is applied to the component of the flow normal to the wavefront yields

$$V_n dV_n = g'dh \tag{6.4.6}$$

which defines the height change that must accompany any acceleration to be consistent with the assumed shallow water flow dynamics. The momentum equation can also be integrated along a streamline to determine the energy balance given by a Bernoulli equation:

$$\frac{V^2}{2} + g'h = constant \tag{6.4.7}$$

which implies that the total energy (kinetic plus potential) must be conserved following a parcel. The implication from this equation is that for high winds, the inversion must be low, and for weak winds, the inversion will be high. This equation and those above govern the structure of the inversion h and the flow speed V. These equations can be integrated to get a solution for a given angle θ. However, we can deduce the characteristic behavior of the flow by simply considering the changes implied by these equations.

Consider first the situation when β and θ are in opposite directions as shown in Fig. 6.10(a), which occurs when the coastline bends away from the flow. In this case, the layer must expand, and there will be flow acceleration downstream as seen from the relationship given in equation 6.4.5.

$$dV_n = -V\frac{d\theta}{cos\beta} > 0. \tag{6.4.8}$$

This is true since $d\theta$ is less than zero. This implied acceleration requires the inversion base to drop as seen by the Bernoulli equation. A higher wind speed and lower inversion base act to increase the Froude number locally. Thus, $sin\beta = \frac{1}{Fr}$ decreases and β gets smaller. This continues until the flow has fully adjusted to the new coastal direction. The region over which β gets smaller results in an expansion fan over which the inversion base drops gradually to its lower base as depicted in Fig. 6.10(a) as the wavefronts rotate toward the coast. Each wavefront represents a lower inversion height from the previous one.

Considering the situation where β and θ are in the same direction as shown in Fig. 6.10(b), which occurs if the coastline bends into the flow, a flow deceleration must occur. This is evident from the equation above since $d\theta > 0$ yielding $dV_n < 0$ or deceleration. As the flow decelerates, the inversion base must go up as implied by the Bernoulli equation. This results in a decrease in the Froude number, which decreases $sin\beta$ thereby increasing β. This implies that the wave fronts should now occur upstream of their original position, which can't occur physically. This results in the wave fronts piling up to produce an abrupt increase in the inversion base height, which becomes a hydraulic jump. The expansion fan and hydraulic jump are analogous to a stream flow piling up against a wall or spreading out as a channel widens. The key result is that this dynamic adjustment process causes height changes in the inversion base near coastal bends. The flow responds to produce higher or lower wind speeds than were observed upstream. The sketch in Fig. 6.10 shows schematically this model of the flow behavior near Pt. Arena under the pattern 3 flow situation examined above, where the inversion height gradually descends in the expansion fan and abruptly goes up in the hydraulic jump.

Certainly, the generality of this result and its frequency of occurrence are an obvious concern. Analysis of the flow along the California coast from experimental periods suggests that the flow is very often supercritical with the inversion below the coastal mountain tops. This effect should, in some sense, be a nearly daily occurrence and model-generated climatologies seem to bear this out. However, a check of the buoy time series of surface winds to the southwest of Pt. Sur suggests that considerable day-to-day variability exists. This variability is probably tied to synoptic-scale changes that are sometimes rather subtle in form. Observations from aircraft flights near Pt. Sur show the horizontal extent and structure of one of these expansion fans. Figure 6.11(a) shows the wind speed distribution, which shows an extensive area of increased winds extending to the southwest of Pt. Sur. This corresponds to lower potential temperature values and a lower inversion height in Fig. 6.11. A plot of the surface pressure (Fig. 6.11(b)) and MBL height (Fig. 6.11(c)) generally agrees with this hydraulic flow analogy, although some important differences do occur. First, the highest winds actually occur downstream of the lowest pressure, and the MBL depth change occurs over a broader scale than the wind enhancement. Concerning the offset in the lowest pressure, evidence suggests that there is an adjustment in the overlying layer that alters the pressure distribution compared to the shallow water model. As

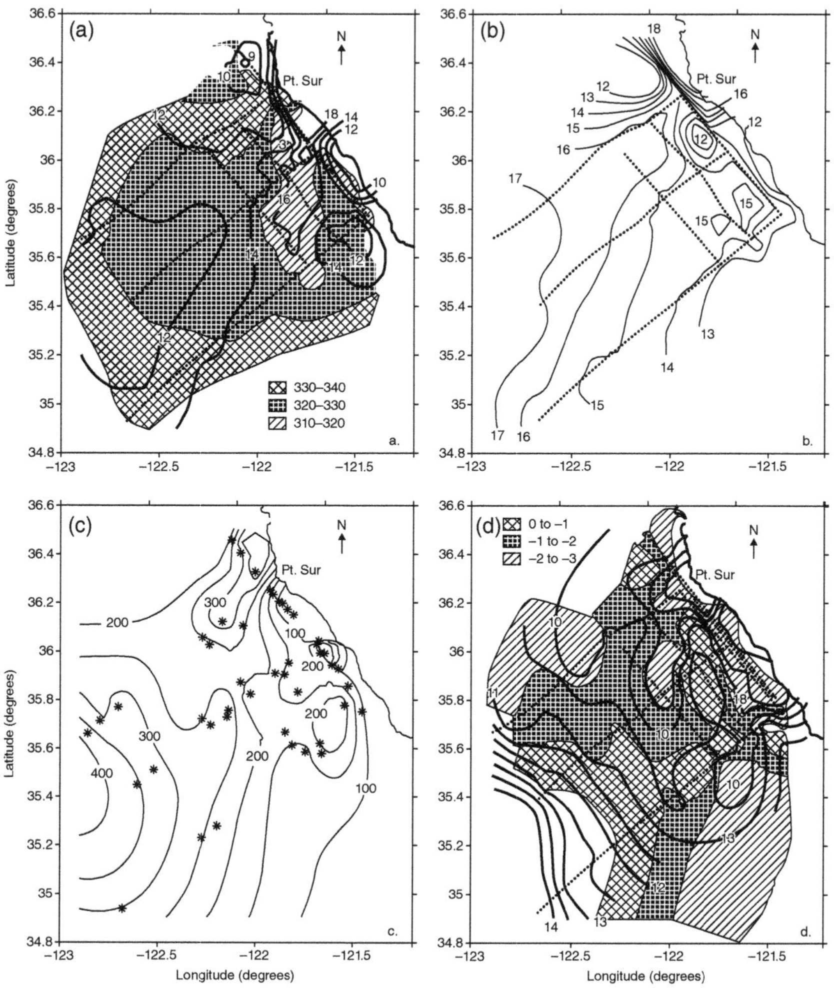

Fig. 6.11 Horizontal structure of (a) winds (solid contours) and potential temperature (hatched regions), (b) pressure, (c) inversion base height, and (d) sea surface temperature along the coast of an observed coastal jet near Pt. Sur California. From C. Dorman et al. (1999). ©American Meteorological Society. Used with permission.

the inversion drops, air above the inversion apparently subsides further, resulting in warming just above the MBL inversion. This warming may contribute to decreasing the pressure before the marine layer has completely shallowed. How this adjustment in the overlying layer might feed back on the basic dynamics has not been examined in previous studies, and certainly the impact of thermal advection and other processes above the MBL have yet to be quantified.

While the shallow water interpretation of supercritical flow adjustment adequately explains the small-scale features that are observed, a more complete description of the full vertical structure also shows a consistent response over a broad range of scales. As the coastline curves away from

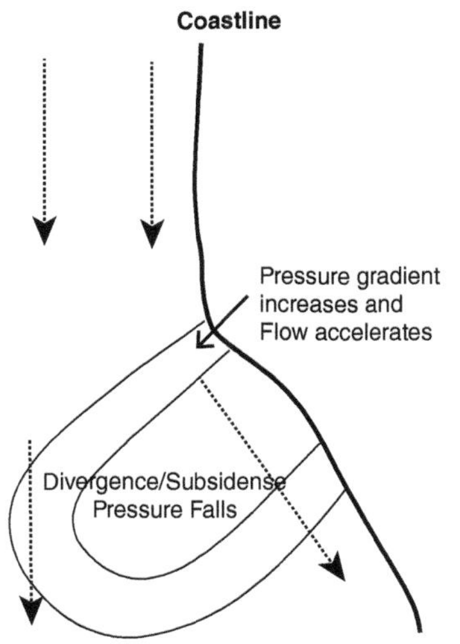

Fig. 6.12 Sketch of larger-scale flow adjustment due to coastal bend. Divergence, subsidence, and low pressure occur south of the point, resulting in flow acceleration near the point and the fastest winds downstream.

the flow south of a point, surface divergence typically occurs, even though the flow may not be supercritical. This surface divergence results in forcing subsidence south of the point, which results in warming above the MBL as well as a drop in the inversion base height. Consequently, a mesoscale low-pressure perturbation occurs just south of the point. The along-shore pressure gradient is subsequently increased near the point to force flow acceleration, causing the highest winds to occur coincident with the lowest pressure. This process is depicted in Fig. 6.12 and explains why some of the same wind enhancements near points and capes seem to occur in coarse resolution models that definitely don't replicate supercritical flow effects.

Numerical model forecasts from mesoscale models seem to produce these features in the wind field with ease. In fact, model forecasts for the case depicted in Figs. 6.11(a)–6.11(d) are quite realistic in their primary features consistent with both shallow water model interpretations as well as the more complete three-dimensional adjustment described above. This suggests that if the MBL depth changes and associated warming above the MBL can be captured, the model will produce realistic expansion fan-type features.

Finally, it is worth noting that coastlines elsewhere than California may also be characterized by these types of supercritical flow effects. The west coast of South America, Southwest Africa, Venezuela, and possibly other locations may possess the basic ingredients for expansion fan effects. These all have reasonably strong low-level inversions and high coastal winds in the vicinity of coastal mountains, which would be conducive to

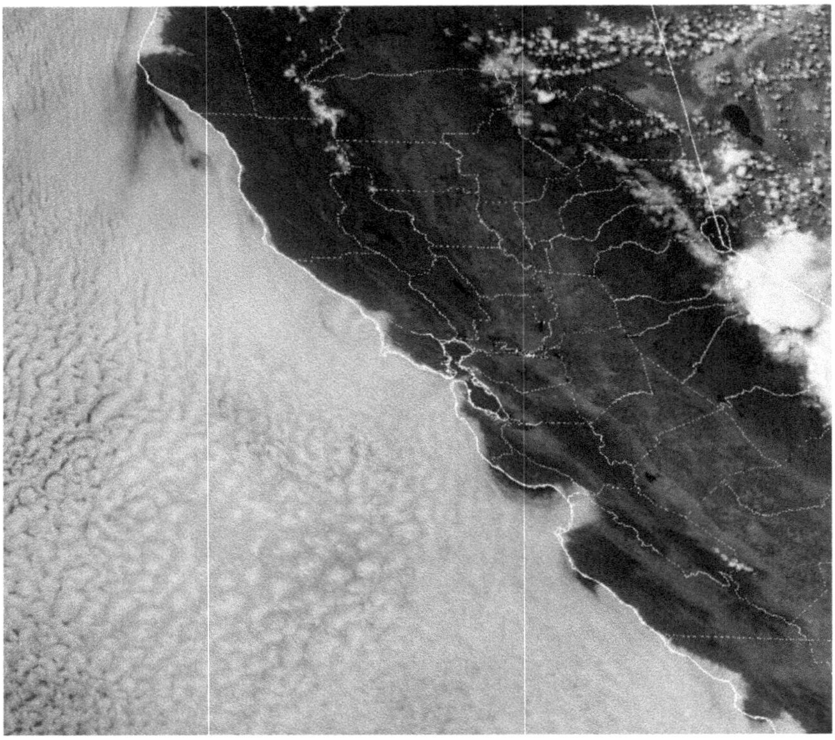

Fig. 6.13 Satellite image showing clearing that occurs south of various points and capes.

this type of flow, although these effects have not been fully documented in other regions. In addition, it is worth noting that these effects can occur on smaller scales with any suitable stretch of coastal mountains and coastal bends. Certainly, a look at the satellite imagery for the California coast in Fig. 6.13 shows clearing near many coastal locations, which hints at numerous small scale features that lower inversion heights where the coastline changes direction. The lower inversion height allows for earlier cloud burn off than the deeper layer offshore and elsewhere along the coast.

6.5 Tip Jets and Other Types of Coastal Jets

Jets other than those associated with a sloping low-level inversion and low-level baroclinic structure can occur. Most notably, flow interacting with coastal topography can be blocked by the topography to force a barrier jet. The basic dynamics of the along-barrier flow due to flow blocking was covered in Chapter 4 on topographic forcing. The occurrence and development of barrier jets in response to landfalling cyclones will be covered in Chapter 11. However, it is worth noting that when air flows around the edge of an island or isolated topographic feature, a low-level jet can form at the topographic tip and extend downwind from the tip. These features are known as tip jets. A well-known example of a tip jet occurs at Cape Farewell at the south tip of Greenland, where strong low-level winds are frequently observed under a variety of flow conditions. The dynamics that create tip jets near Greenland have been characterized as several types. One type, as documented by Doyle and Shapiro (1999), occurs with flow over the topography. In this situation, the downslope flow on the lee side accelerates to produce a strong surface flow in the lee. In this scenario, the flow is from the west with sufficiently weak stratification to allow the flow to cross the barrier. The tip jet forms to the south of a lee trough where downslope flow occurs. The lee trough sets up a pressure/thermal gradient to which the flow geostrophically adjusts downstream to produce the jet. Another type of tip jet forms when the low-level flow toward Greenland is blocked by strong low-level stability, and the flow accelerates around the tip toward low pressure on the lee side. This scenario is also characterized by westerly flow toward Greenland, which results in a jet that extends east-southeast of the tip of Greenland as shown by Moore and Renfrew (2005). The position of the tip jet can change based on the relative angle of the flow toward the barrier. Similar features occur around many islands with sufficient topography to block the low-level flow. Moore and Renfrew (2005) also noted a third type of tip jet that is described as a reverse tip jet. In this scenario, there is easterly or northeasterly flow toward Greenland with high pressure to the north that produces anticyclonic flow. The flow east of Greenland may be blocked to produce low-level flow acceleration to the south toward the tip. The flow turns anticyclonically at the tip in essentially gradient wind balance. This gradient flow is supergeostrophic due to the anticyclonic turning.

The formation and character of tip jets can be illustrated by considering an example of the flow as it interacts with Greenland. Figure 6.14 shows the synoptic scale structure with high pressure over Greenland and low pressure south of Iceland. The example generally shows large-scale easterly flow impinging on Greenland, which sets up a reverse tip jet type flow. A region of high winds forms on the southern tip as it turns anticyclonically. Recall that the flow around and over topography depends upon low-level stratification and incident flow speeds. In order for flow to be forced around the barrier, sufficient low-level stratification needs to occur to prevent flow over the topography. This sets up a lee vortex and a strong pressure gradient that extends out from the tip. As can be seen in the cross-section in Fig. 6.15, where relatively strong stability occurs in the low levels and the inversion slopes up toward the coast. This upward slope in the isentropes is consistent with the thermal wind acting to increase the easterly flow down toward the surface.

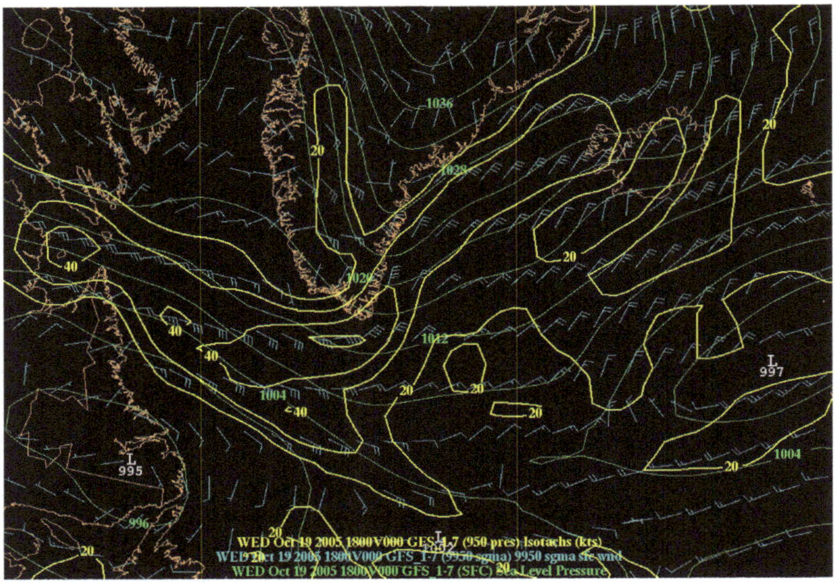

Fig. 6.14 Horizontal structure of a tip jet at the south tip of Greenland. Isotachs are solid yellow lines, sea-level pressure is green, and wind barbs at the surface are shown.

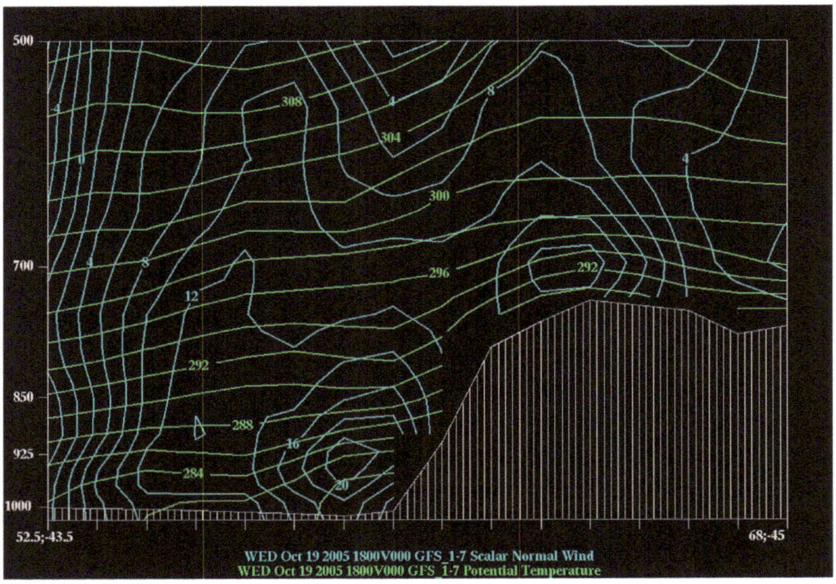

Fig. 6.15 Cross-section through the Greenland tip jet. Isotachs are cyan contours and isentropes are green contours. The isentropes tilt down offshore to produce a cross-coast thermal (pressure) gradient consistent with thermal wind balance and the wind increase toward the surface.

6.6 Exercises

6.1 Explain how coastal jets owe their existence to low-level thermal wind balance. Based on this idea, how would the strength of the flow at 850 hPa impact the strength of the coastal jet.

6.2 Calculate the magnitude of the coastal jet wind speed assuming a cross-coast thermal difference of 5°C over a 200 km distance and averaged through the surface to 850 hPa layer. Assume the 850 hPa wind is zero.

6.3 The sloping inversion is typically the main contributor to the cross-coast thermal gradient that leads to the coastal jet. Calculate using the thermal wind equation, the jet wind speed if the inversion slopes from 1500 m down to 200 m over a 200 km distance. Assume the inversion strength is 12°C and the wind above the inversion is zero. What would the jet speed be if the inversion strengthens to 20°C?

6.4 Coastal jets occur along other coastlines besides the US West Coast described in the text. Some of these include the west coast of South America, the coast of Oman and Yemen, and the coast of South Texas and Northern Mexico. Assuming that thermal wind balance and low-level flow blocking by coastal topography are necessary ingredients for a coastal jet, explain the synoptic conditions that occur in these regions to cause a coastal jet. Note the location of high and low pressure relative to the coast and which direction the coastal jet blows.

6.5 Given that the flow along the coast is often characterized as supercritical, calculate the height change, wind speed increase, and expansion fan orientation downstream from a bend in the coastal topography. Assume the upstream wind speed is 18 ms^{-1}, the inversion height is 800 m, strength is 10°C, and that the inversion height drops to 300 m.

6.6 The coastal jet tends to be strongest on days when the isobars are at an angle to the coast. Explain why the jet is stronger when the pressure gradient is at an angle to the coast as opposed to directly across the coast. If the surface geostrophic flow is 10 ms^{-1} offshore and it makes an angle of 20° to the coast, calculate the magnitude of the along-coast pressure gradient. Calculate the speed increase that would occur over a 100 km distance along the coast.

7 Coastally Trapped Wind Reversals

Coastally trapped wind reversals occur along mountainous coastlines under conditions that produce a capping inversion below the height of the coastal topography. These types of events have been most studied along the US West Coast and are commonly referred to as stratus surges or southerly surges due to the characteristic fog or marine stratus and southerly winds that propagate northward up the coast. The forecasting of these events is important due to their impact on coastal airports and marine activities where visibility is an issue. The switch from northerly to southerly winds may also be important in modifying the prevalent summer-time coastal upwelling. Coastally trapped wind reversals, while most studied along the US West Coast, occur in other regions around the world as well. There is a direct counterpart along the South American West Coast and they have been documented along the South African coast, the north coast of Spain, and the Arctic coast of Alaska.

7.1 Description of Wind Reversals

Coastally trapped wind reversals are characterized by an abrupt change in the direction of the along-coast flow at a particular location. The characteristic time series evolution of observed variables at coastal buoy locations has been studied by Bond, Mass, and Overland (1996) and is depicted in Fig. 7.1. The wind reversal is characterized by a rise in sea-level pressure, a wind reversal in the along-shore wind component, little temperature change, and typically a change from clear to cloudy conditions. To the extent that hourly buoy observations resolve these events, the wind reversal appears to be nearly instantaneous, and an examination of the time of wind shift at progressively further north buoys shows that the wind shift propagates to the north along the US West Coast. This northward propagation is shown in Fig. 7.2 that shows the mean time of wind reversal relative to the wind shift at a buoy along the California coast just north of San Francisco (buoy 46013). Typically, the winds reverse at later times further to the north and earlier to the south of this location. An estimate of the speed of propagation can be obtained from the slope of the line connecting these times of wind reversal. The typical propagation is around

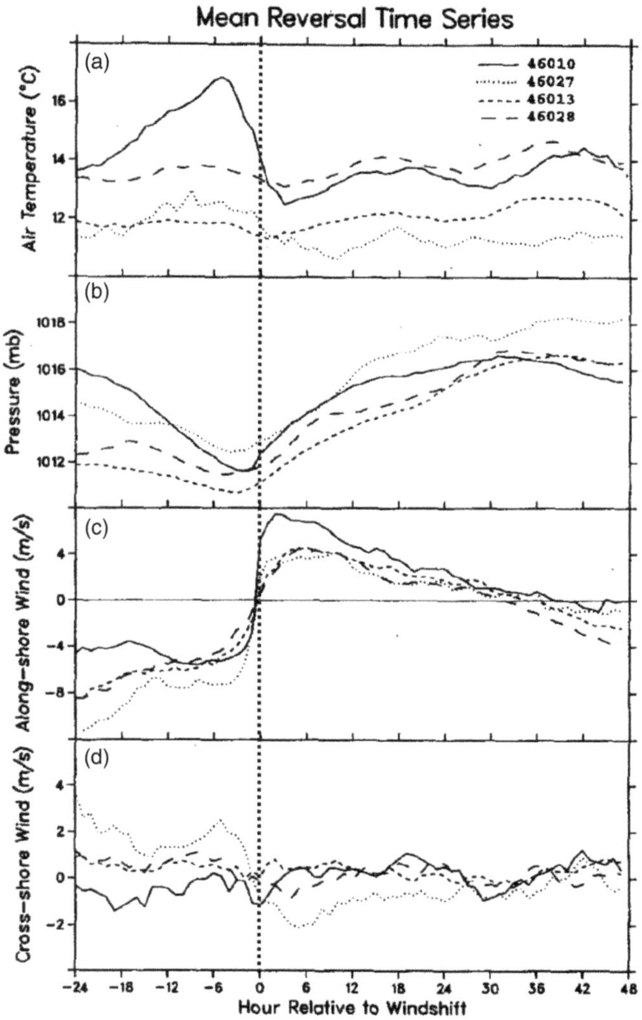

Fig. 7.1

Composite time series of (a) air temperature, (b) pressure, (c) along-coast wind, and (d) cross-coast wind for coastal wind reversals at four buoy locations from central California to Washington. From Bond, Mass, and Overland (1996). ©American Meteorological Society. Used with permission.

6 m/s based on the composites from Bond, Mass, and Overland (1996) as well as individual cases such as Dorman (1985). This propagation speed is close to the estimated linear Kelvin wave propagation speed as noted by Dorman (1985). Based on this correspondence, these disturbances have been characterized as coastally trapped Kelvin waves (Dorman (1985) and others).

Alternative explanations of these wind reversals are that they are gravity currents, which is suggested by the abrupt reversal and rapid temperature drop accompanying some events. This explanation from Mass and Albright (1987) is supported by time series of the vertical structure of

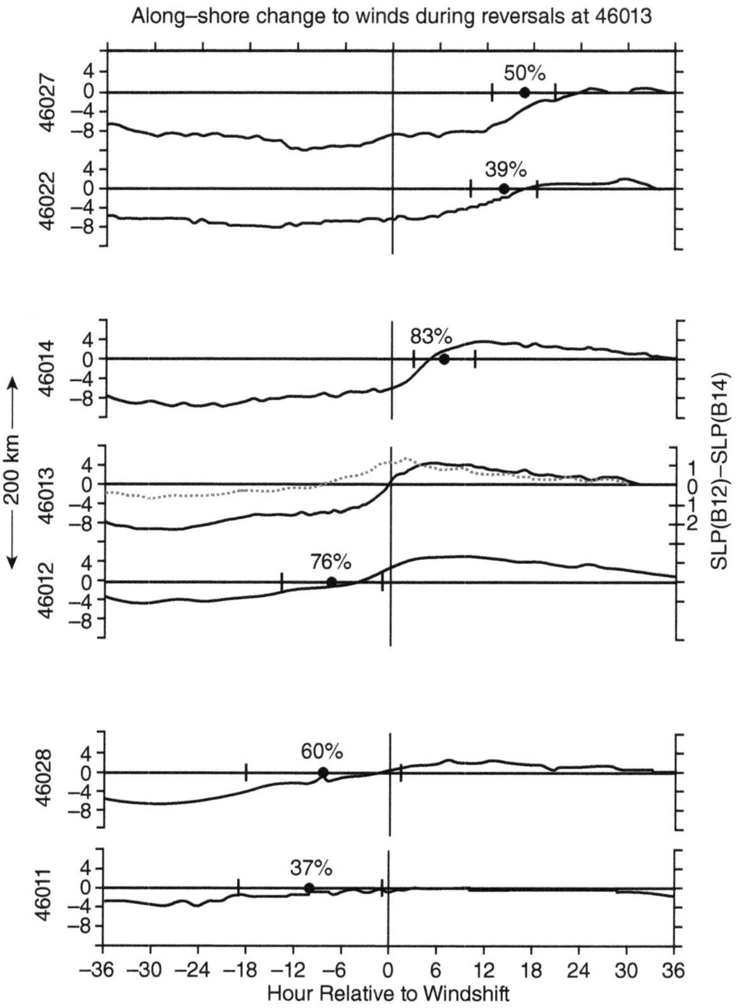

Fig. 7.2 Time series of along-coast wind component for buoy locations along the California coast. Time of wind reversal at buoy 46013 is used as reference (0 hour) and spacing on page of plots is proportional to along-coast distance. From Bond, Mass, and Overland (1996). ©American Meteorological Society. Used with permission.

one of these disturbances that was obtained using a 915 MHz wind profiler located at Ft. Ord near the Monterey Bay. Figure 7.3 shows a time-height cross-section that indicates an abrupt deepening of the marine layer and a drop in temperature with the passage of the disturbance. The time series is highly suggestive of a gravity current with a well-defined gravity current head. Although this structure is suggestive of a gravity current, the hourly temporal resolution of the profiler and its location inland from the main coastline do not preclude other explanations including that of a Kelvin wave.

The vertical structure does, however, clearly show that after the passage of this disturbance, the marine layer is deeper and the southerly winds

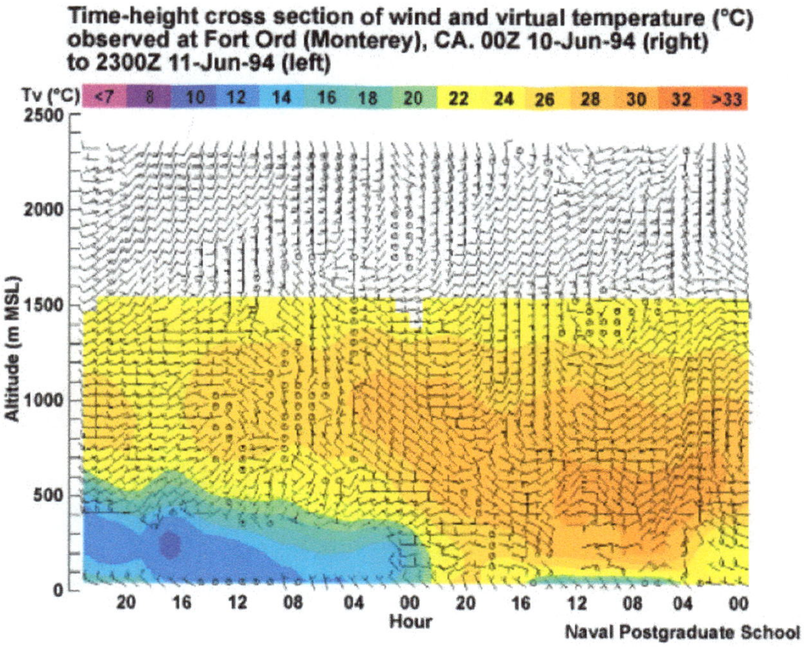

Fig. 7.3

Time-height series from a wind profiler at Ft. Ord, California on June 10–11, 1994 when a coastal wind reversal occurred. Virtual temperature is the color fill with the cool marine layer temperatures shown as blue. The wind reversal passed just offshore of the site at about 2200 UTC, which corresponds to the increase in (development of) a marine layer.

occur not only in the marine layer but above as well. The deeper marine layer is consistent with the clouds that accompany the reversal. Cooling associated with the deepening of the layer is responsible for the formation of the clouds.

The horizontal structure of a coastally trapped wind reversal reveals the cross-coast length scale and relationship between southerly winds and stratus. Figure 7.4 shows a satellite image with the available surface observations plotted. The image indicates a narrow band of stratus along the coast that is about 100 km wide. The surface wind observations suggest that the southerly winds are correspondent with the cloudy regions. Analyses of sea-level pressure show that the cloudy region is also characterized by higher pressure, which indicates the presence of a deeper marine layer. The deeper marine layer is consistent with marine stratus and the time series observations from the profiler examined above. The presence of the deeper marine layer and associated higher pressure supports a south-to-north pressure gradient along the coast and southerly winds. A crucial question concerning these disturbances is whether the deeper marine layer is a response to external forcing and whether it is an active driver of the propagation as suggested by Kelvin wave theory. A question about the dynamics of these reversals concerns the importance of the marine layer structure versus other factors above the marine layer in establishing the

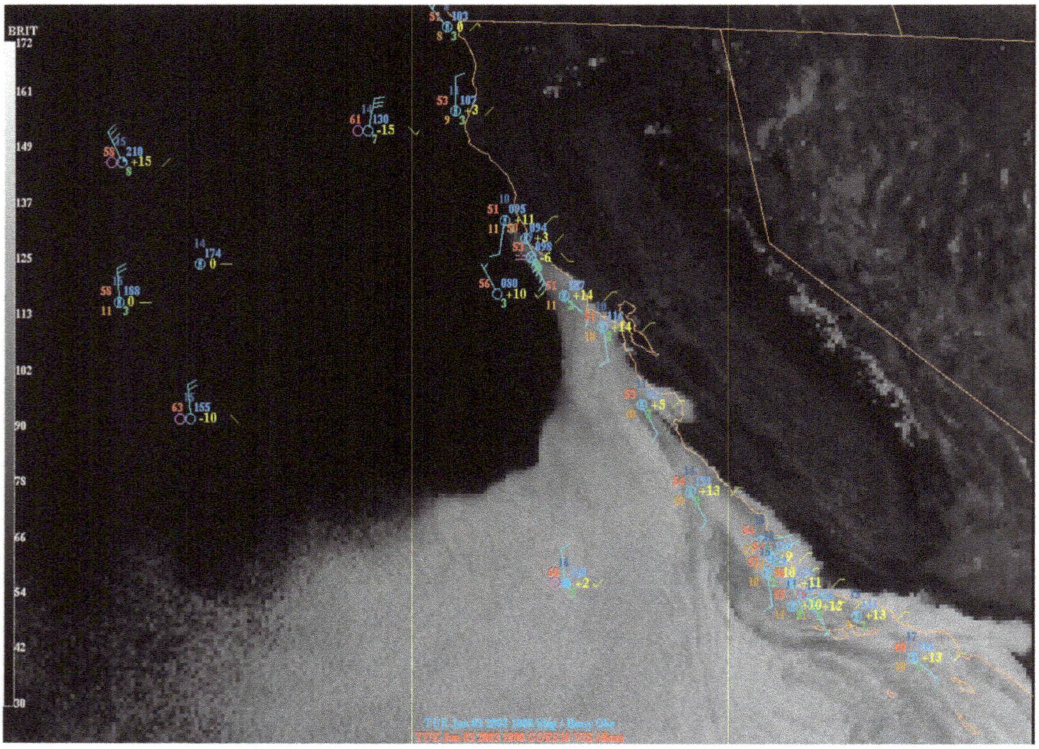

Fig. 7.4 Visible satellite image with buoy observations showing the relationship of clouds and the occurrence of southerly winds. Winds offshore in the clear air are from the northwest, and near shore in the cloudy region they are from the southeast.

sea-level pressure gradient that leads to southerly winds along the coast. Another possible explanation about the cause of coastally trapped wind reversals is that they arise because the synoptic scale pressure gradient reverses, and this leads to trapped ageostrophic flow toward the north where the lowest pressure along the coast is found. This explanation does not depend upon any particular marine layer structure as long as the larger-scale processes are sufficient to result in a region of locally lower pressure to the north along the coast. Section 7.2 will consider this scenario in greater detail as the synoptic structure is examined more completely.

7.2 Synoptic-Scale Forcing of Reversals

Mass and Bond (1996) have documented the characteristic large-scale structure and its evolution that is associated with coastal wind reversals along the North American coast. The basic structure of the large-scale atmosphere at the time of wind reversal is depicted in Fig. 7.5.

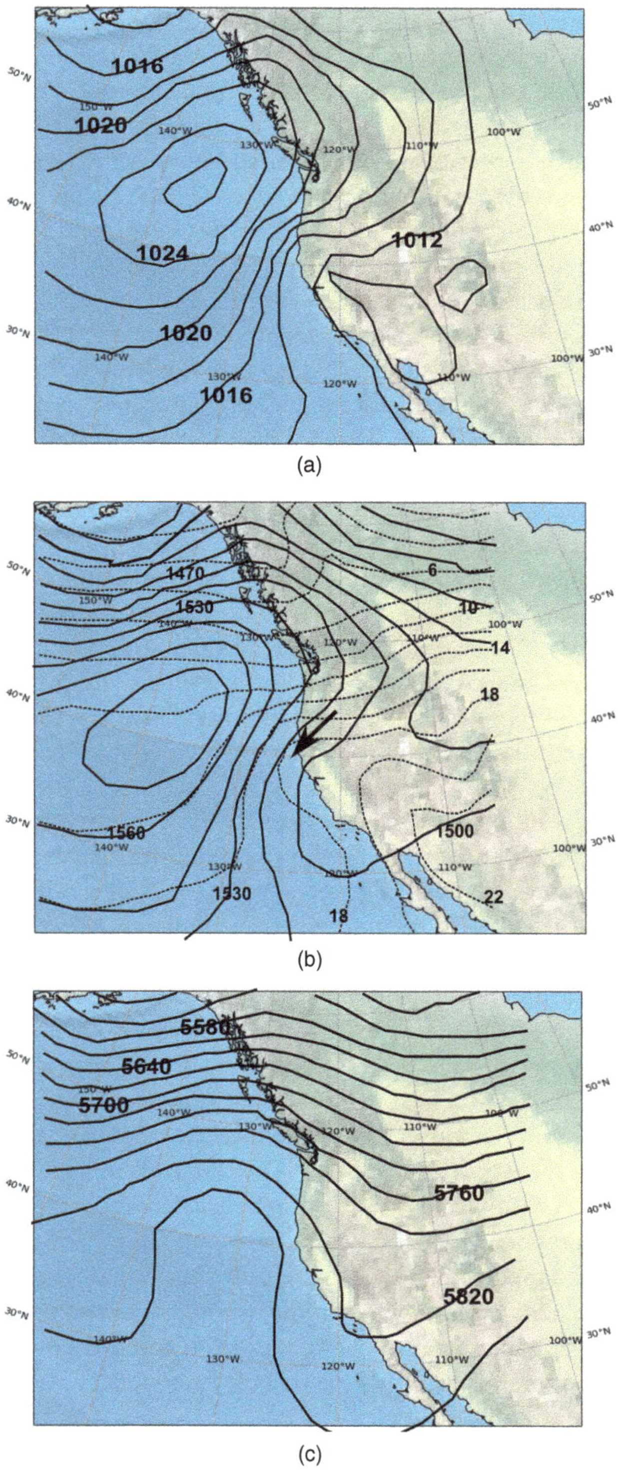

Fig. 7.5 Composite synoptic-scale patterns at the time of a coastal wind reversal at Buoy 46013 for (a) sea-level pressure, (b) 850 hPa height (solid lines), temperature (dashed lines), and sample wind over Northern California, and (c) 500 hPa height.

The sea-level pressure analysis shows that the axis of the thermal trough, normally located through the central valley of California, is shifted toward the coast and just offshore in northern California. This establishes the south-to-north pressure gradient to produce trapped southerly flow. Another crucial aspect of the low-level structure is the presence of surface ridging into the Pacific Northwest which results in a pressure gradient from northeast to southwest across southern Oregon and northern California. This pressure gradient supports low-level offshore directed flow. This structure is most evident in the 850 hPa analyses in Fig. 7.5(b). The 850 hPa structure is similar to the sea-level pressure and clearly depicts a region of geostrophic northeasterly flow across northern California. The observed 850 hPa winds also show this offshore flow, and the thermal analysis shows this region of offshore flow to be characterized by warm temperatures. The warm temperatures, or thermal ridge, are coincident with the lower sea-level pressures clearly showing the importance of the thermal structure above the marine layer in reversing the synoptic scale pressure gradient. Presumably, this thermal ridge along the coast is associated with down-slope flow and lee troughing along the coastal mountains in northern California. Looking at the 500 hPa structure in Fig. 7.5(c), the low-level ridging over the Pacific Northwest occurs downstream from the 500 hPa ridge axis. This suggests that synoptic-scale subsidence forced from aloft is contributing to raising the surface pressures over the Pacific Northwest.

The time evolution of these basic synoptic-scale features was also examined by Mass and Bond (1996) and shows a very consistent evolution of the large-scale features. At 500 hPa, the upper-level ridge amplifies offshore of the West Coast 48–36 hours prior to the initiation of a wind reversal as shown in Fig. 7.6. The amplifying ridge reaches maximum amplitude just offshore around the time of reversal and moves onshore just after the reversal is started. Over the next 24–48 hours, this ridge moves onshore and weakens as does the 850 hPa and surface ridging over the Pacific Northwest. The decay of the surface and 850 hPa ridge result in diminished offshore flow and the movement of the thermal ridge back onshore. The weakening of these low-level features is also associated with the return to northwesterly coastal winds as the pressure gradient returns to a north-to-south configuration. This basic synoptic evolution of the sea-level pressure gradient does support the idea that the near-shore wind reversal could be an ageostrophic trapped flow down the synoptic scale pressure gradient. The propagation and detailed evolution of this trapped flow is not clear from synoptic composites nor is the role of the boundary layer depth and its evolution.

Idealized modeling studies using these basic synoptic conditions by Skamarock, Rotunno, and Klemp (1999) suggest that a forced Kelvin wave response occurs that may be responsible for the wind reversal and its northward propagation. In this model, which is examined in more detail later, the synoptic scale pattern simply sets up an along-coast profile

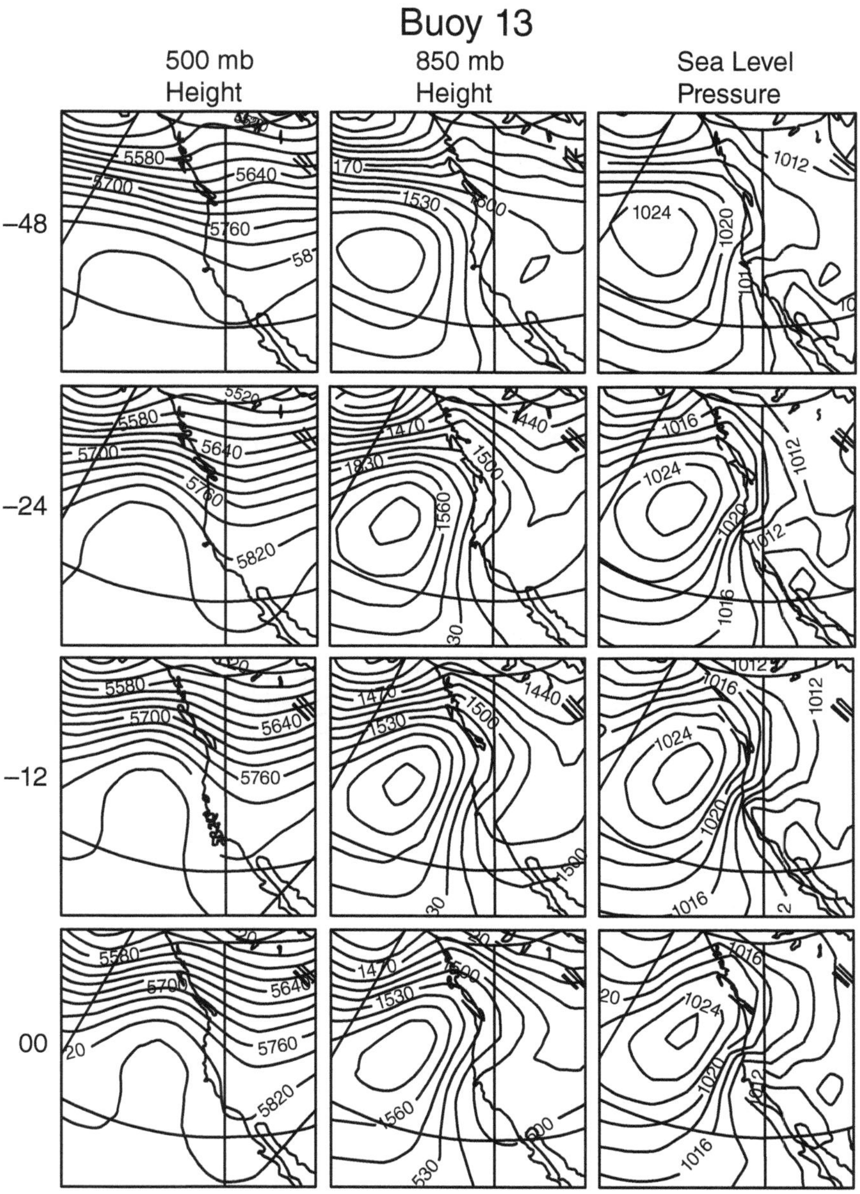

Fig. 7.6 Synoptic scale evolution at 500 hPa, 850 hPa, and sea-level from 48 h prior to 48 hour after a wind reversal at Buoy 46013. From Mass and Bond (1996). ©American Meteorological Society. Used with permission.

of offshore-directed flow near mountain-top level. The offshore flow initially scours out the marine layer, lowering the inversion near the coast, and resulting in lowered surface pressure as a consequence. The region of lowest surface pressure occurs near the region of strongest offshore flow and causes the climatological strong northerly winds to shift seaward north of this feature, and then the flow turns back toward the coast to the

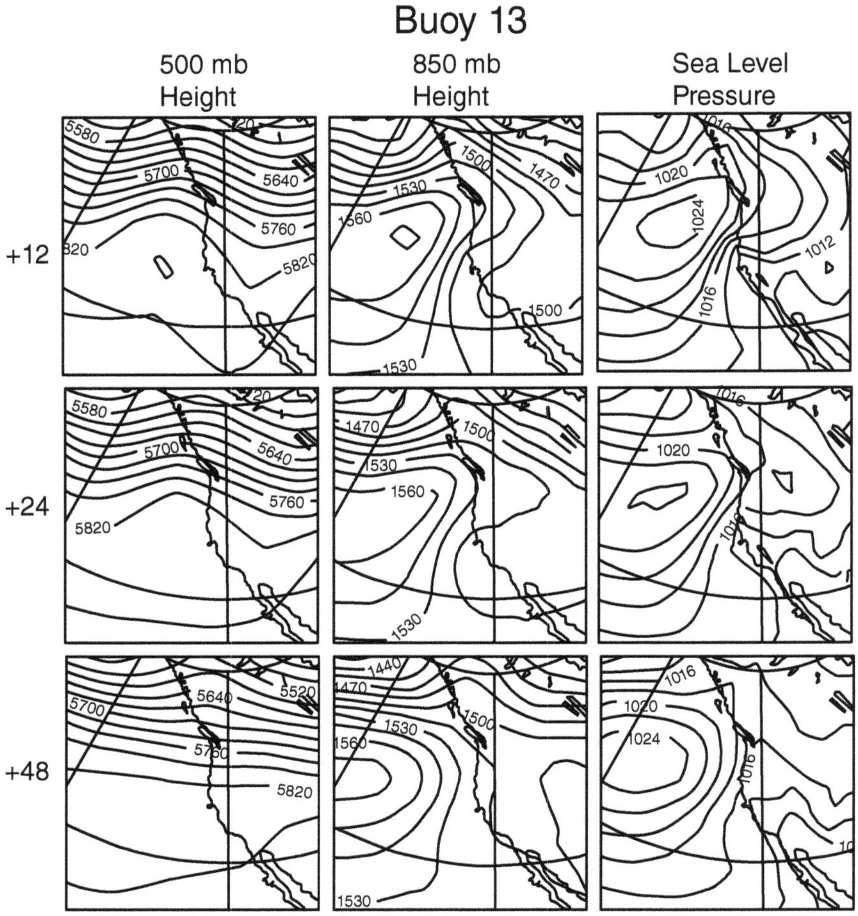

Fig. 7.6 (Cont.)

south of this coastal low. The flow to the south is blocked by the topography and results in elevating the marine layer to spawn a Kelvin wave that subsequently propagates north. This is illustrated later in this chapter to demonstrate the strong coupling between the synoptic scale forcing and the Kelvin wave dynamics that result in a propagating disturbance.

There is evidence that some coastal wind reversals do not propagate while others do (Nuss (2007)). This behavior suggests multiple explanations for these different kinds of events. Nuss (2007) has examined the synoptic conditions associated with these differing coastal responses. While the synoptic-scale patterns for both propagating and non-propagating wind reversals are somewhat similar, the character of the offshore flow is different. For propagating disturbances, the offshore flow was found to be stronger and more localized, similar to the profile used by Skamarock, Rotunno, and Klemp (1999). However, for non-propagating reversals, the offshore flow was weaker and more uniformly distributed along the coast. This broad region of offshore flow produces coastal troughing over much of the coast, within which the winds reverse due to small variations in pressure. Without a localized low, no significant along-coast variations in marine boundary layer depth are forced to spawn a Kelvin wave. Consequently, the subtle detail in the general synoptic pattern seems to be important in determining the exact nature of a wind reversal.

7.3 Character of the Disturbance

Detailed observations of coastally trapped disturbances are typically not available, and knowledge of their structure comes primarily through inferences based on surface observations, and more recently profiler time series and aircraft observations. Time series of buoy observations along the coast were composited by Bond, Mass, and Overland (1996) and show the characteristic evolution of these disturbances. The most telling signature is the rapid wind direction change that occurs as the disturbance passes. The northerly winds along the coast decrease rapidly and then pick up rapidly from the south. Associated with this wind shift, the pressure increases as the disturbance passes. While some buoys suggest that the temperature occasionally drops after passage, in general the temperature does not change much across the disturbance. This is consistent with a Kelvin wave interpretation for these disturbances and not consistent with a gravity current interpretation where the potential temperature would change. A marine boundary layer that is deeper in the disturbance where higher pressure is observed describes a simple Kelvin wave structure. The southerly winds arise as a geostrophic response to this pressure pattern, which propagates up the coast. This structure represents a simple two-layer linear Kelvin wave model where the depth of the lower dense fluid determines the dynamic response.

Detailed observations using instrumented aircraft on a single disturbance in 1994 (Fig. 7.7) by Ralph et al. (1998) show that the structure of these disturbances is more complex than a simple two-layer linear Kelvin wave. The depth of the marine boundary layer along the coast through the disturbance is depicted in Fig. 7.7, which shows a marine boundary layer with a nearly constant depth of 250 m across the leading edge of this disturbance. The thickness of the marine layer inversion, however, changes across the disturbance, expanding to the south. This structure is consistent with the increase in sea-level pressure that occurs across the disturbance. The wind reversal is shown to occur within the inversion layer as it thickens, and this occurs well ahead of the surface wind reversal. The cross-coast structure shown in Fig. 7.8 is consistent with the disturbance in the inversion layer. The marine boundary layer depth does not change appreciably in the cross-coast direction. Instead, the inversion expands toward the coast to produce higher pressure along the coast. The southerly

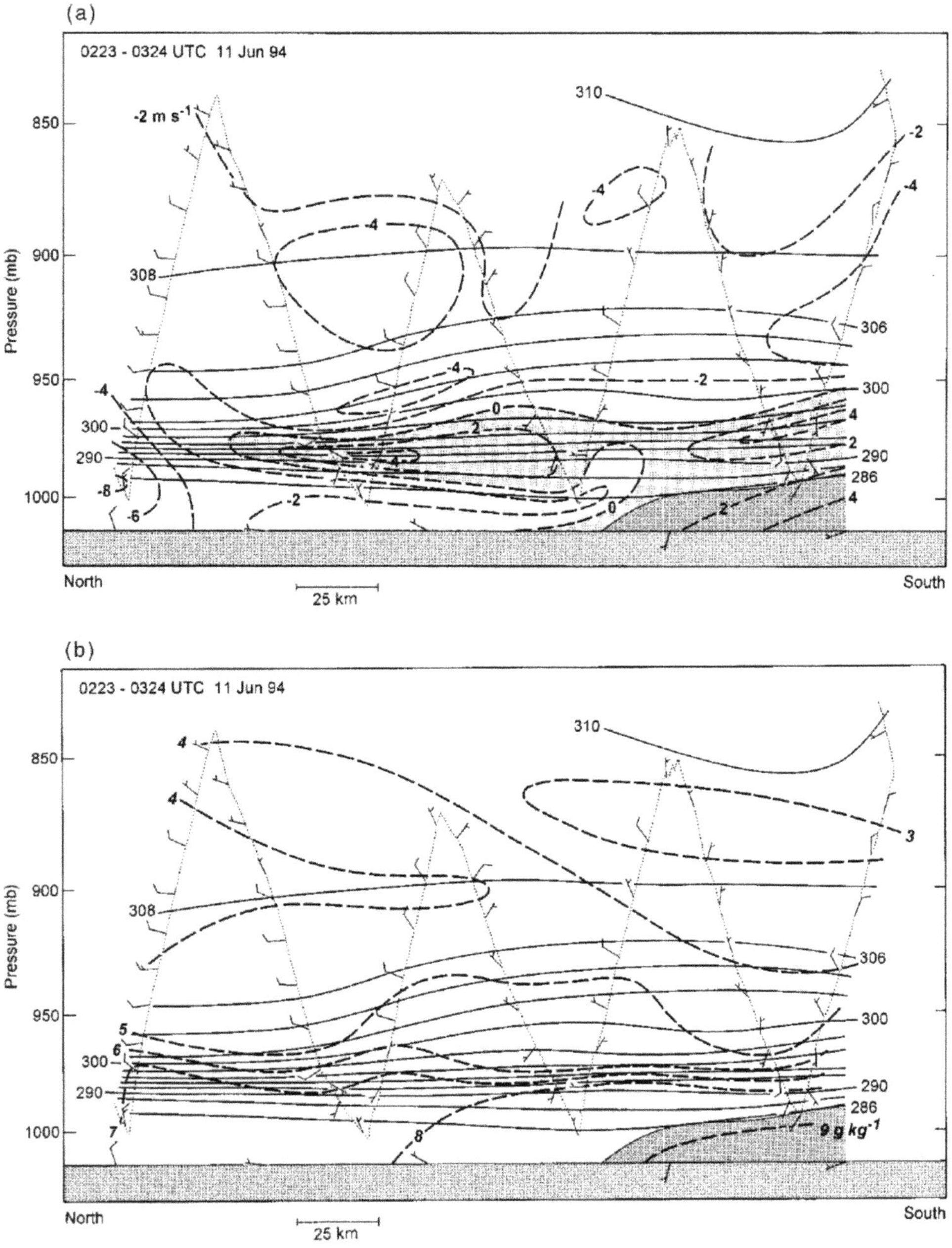

Fig. 7.7 Along-coast cross-section from aircraft measurements of the (a) potential temperature (solid) and isotachs (dashed) and (b) potential temperature (solid) and mixing ratio (dashed). Region of southerly winds is lightly shaded and clouds are darker shaded. From Ralph et al. (1998). ©American Meteorological Society. Used with permission.

winds occur within the expanded inversion layer. This structure suggests a Kelvin wave propagating within the inversion layer instead of a simple two-layer model.

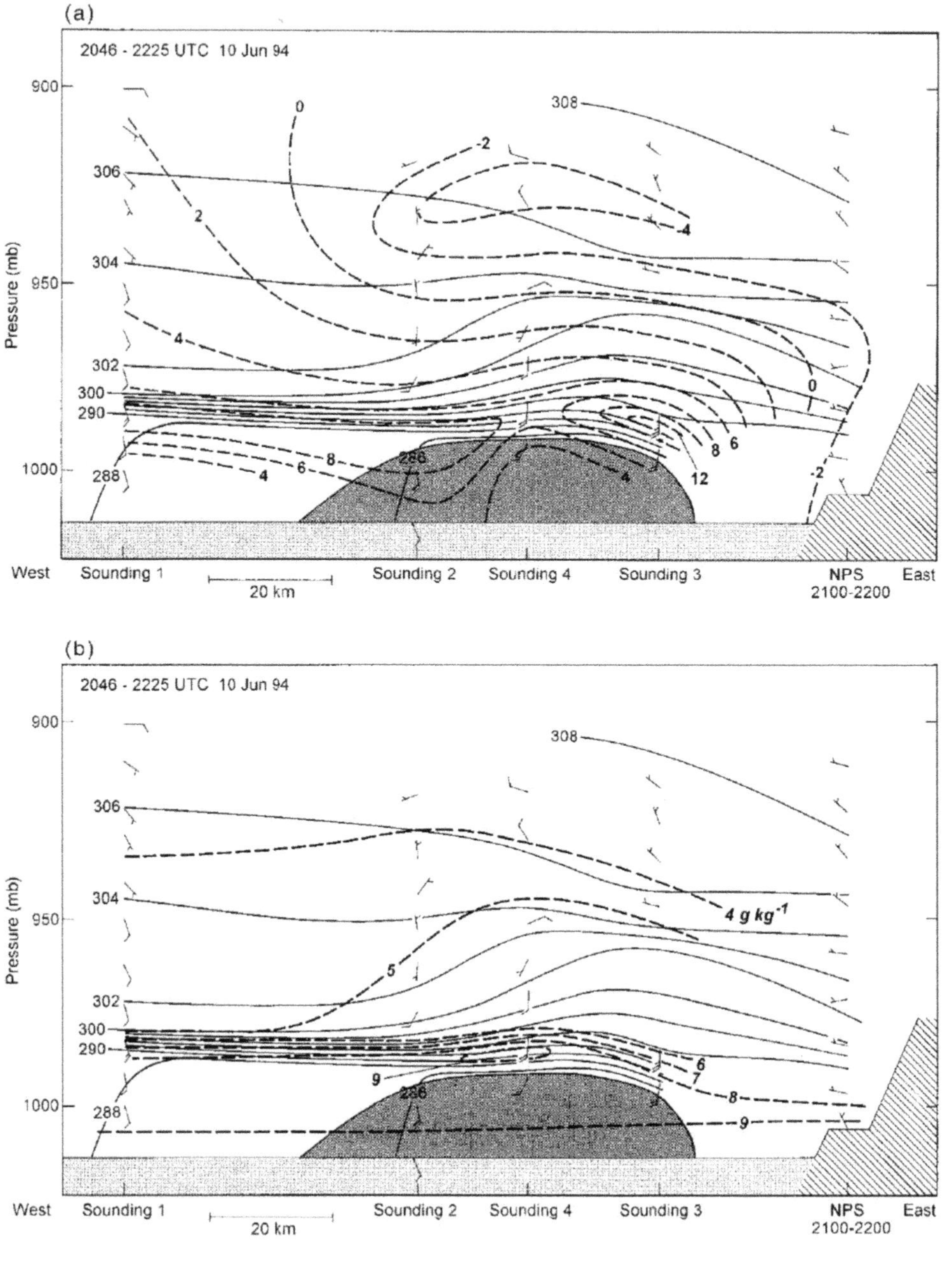

Fig. 7.8 Cross-coast cross-section from aircraft measurements of the (a) potential temperature (solid) and isotachs (dashed) and (b) potential temperature (solid) and mixing ratio (dashed). Region of clouds is shaded. From Ralph et al. (1998). ©American Meteorological Society. Used with permission.

7.4 Conceptual Models of CTWR Development

As suggested in Section 7.1, CTWRs are characterized by boundary layer structures that are consistent with coastally trapped Kelvin waves. The simplest model of a Kelvin wave is that from the shallow water equations associated with a two-layer fluid system. While the observed structure is a bit more complex than two layers, the shallow water Kelvin wave provides some basic insight into the evolution of these events. The propagating disturbance represents a deformation of the fluid interface as illustrated in Fig. 7.9, which is lifted toward the coast and back into the disturbance. This structure is highly consistent with the observed pressure distribution in CTWRs and is consistent with the elevation of the inversion top (if not the base) as found in the observations. Kelvin waves can freely propagate at a speed of about 6–12 m/s depending upon the inversion depth. The propagation will occur as long as the environment is favorable. A favorable environment is simply that the large-scale, along-coast pressure gradient (north to south) is not so large to prevent the perturbation pressure to locally produce a pressure gradient (within the disturbance) from the opposite direction (south to north). For example, if the perturbation pressure gradient across the Kelvin wave front is 0.1 hPa/km, then the opposing pressure gradient must exceed this magnitude to stop the propagation. The large-scale pressure gradient in the along-coast direction might be 10 hPa over 200 km, which is insufficient to halt propagation. This weak pressure gradient occurs along much of the coast, even during northerly flow patterns, except at the north end of the thermal trough which often

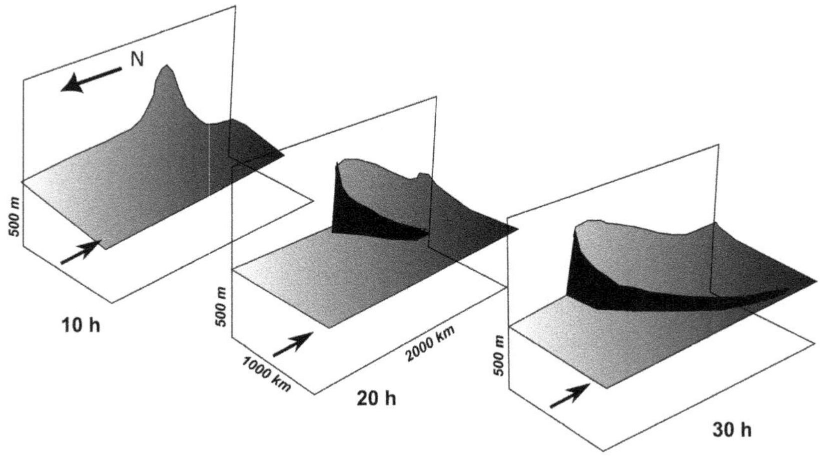

Fig. 7.9 Evolution of the interface in a shallow-water simulation of a Kelvin wave initiated with a localized pressure anomaly.

produces a large north-to-south pressure gradient near the Oregon border. However, CTWRs and Kelvin waves do not occur that frequently due to the need for something to force them to develop. The synoptic-scale provides that forcing, as described above, to generate a forced Kelvin wave initially, as opposed to a freely propagating wave. This interaction between the synoptic-scale forcing and the generation of a Kelvin wave is crucial and has been described by Skamarock, Rotunno, and Klemp (1999).

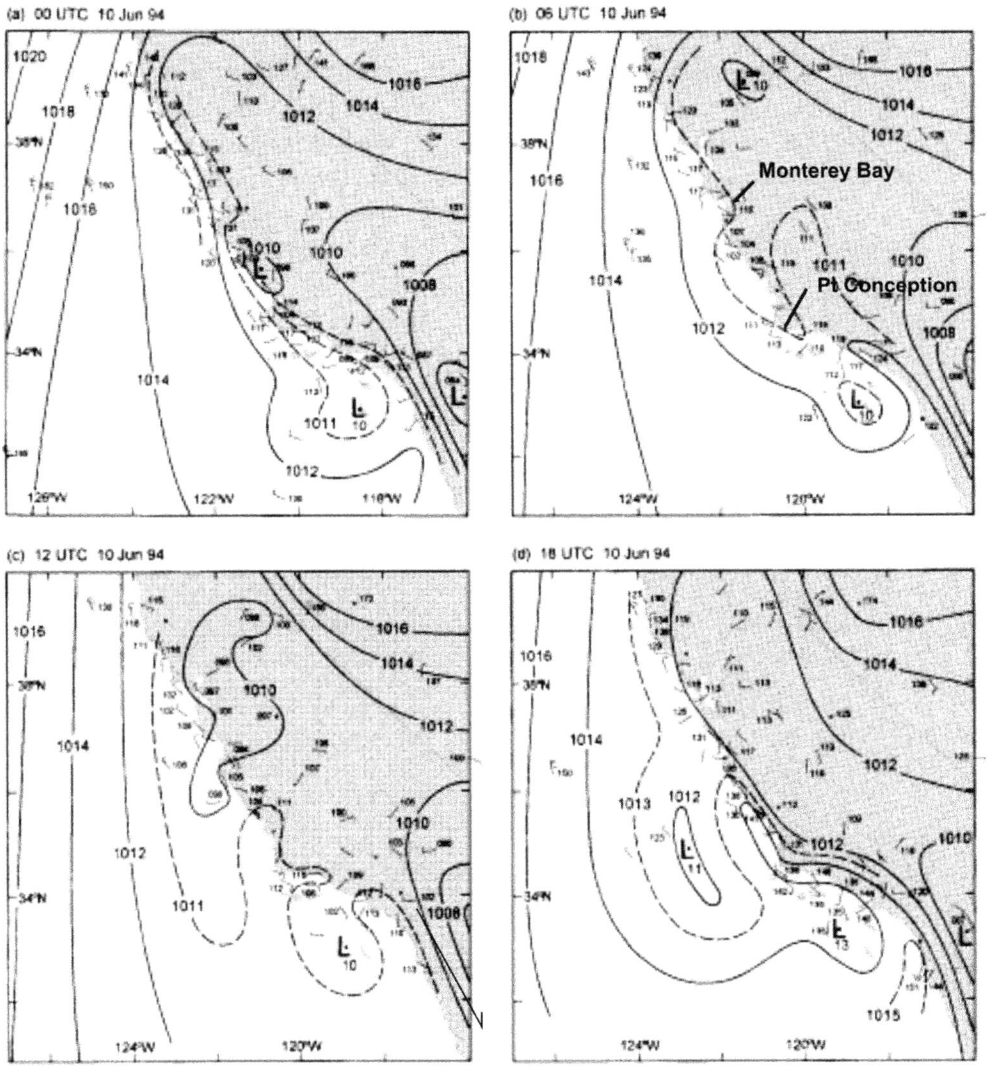

Fig. 7.10 Sea-level pressure analysis over the initial stages of a wind reversal. Localized low pressure develops at a location along the coast where the strongest offshore flow occurs. The low center migrates further offshore overtime as the winds reverse along the coast. From Ralph et al. (1998). ©American Meteorological Society. Used with permission.

To illustrate this interaction, let's consider the evolution of the June 1994 case in a bit more detail. Figure 7.10 illustrates the detailed sea-level pressure analyses every 6 hours during the time when this wind reversal initiates. The sea-level pressure distribution at 0000 UTC on June 10 (Fig. 7.10(a)) is prior to the wind reversal and shows a relatively well-defined cross-coast pressure gradient with a weak low in the Southern California Bight area. Through the next 12 hours, a region of coastal low pressure develops near Monterey Bay and just south of it (Fig. 7.10(b) and (c)). This location corresponds to the strongest offshore flow. By 1200 UTC, the wind reversal has started near Pt. Conception due to the reversed pressure gradient to the south of this coastal low. As the CTWR develops (Fig. 7.10(d)), a narrow ridge of high pressure along the coast develops in association with the clouds and southerly winds. This feature propagates north over time and has a structure that is consistent with a Kelvin wave. It is useful to note that the region of low pressure along the coast does not propagate north but strengthens in place off the coast along the axis of maximum offshore flow. Consequently, there is little evidence that this coastal low is propagating due to synoptic scale adjustment.

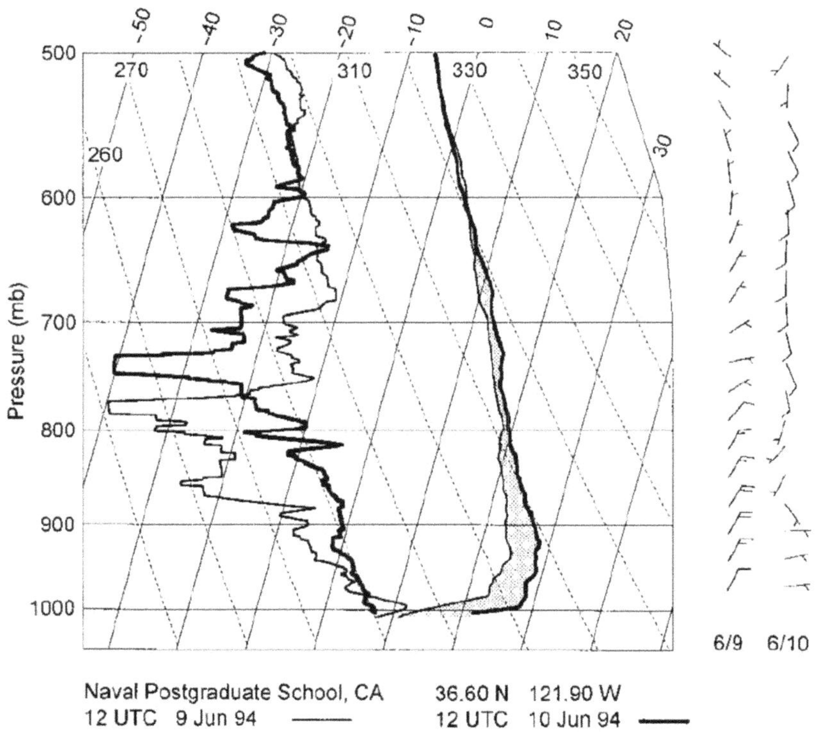

Fig. 7.11 Soundings taken one day prior and day of a coastal wind reversal along the coast south of Monterey, CA. Shaded region represents the low-level warming that produced lower surface pressure at this location. From Ralph et al. (1998). ©American Meteorological Society. Used with permission.

An important question is how the synoptic scale forces this coastal low. Mass and Bond (1996) hypothesized that it was downslope warming associated with lee troughing. While this may occur in some cases, analysis for this case shows that offshore advection of warm air above the boundary layer was almost entirely responsible for the observed pressure change associated with the coastal low. This is illustrated by the comparison of two soundings taken near Monterey at the Naval Postgraduate School (NPS) on June 9 and June 10 at 1200 UTC shown in Fig. 7.11. The soundings show strong low-level warming below 700 hPa and mostly below 800 hPa. While the winds on June 9 are definitely offshore, an analysis over the broad region allowed the computation of temperature advection and shows this warming to be due to warm advection and not subsidence. This dependence on warm advection may account for the strong seasonal dependence of coastal wind reversals, which rarely occur in the cool season. The synoptic forcing for this event showed offshore flow across the central California coast that produced localized low pressure along the coast, which is similar to the explanation given by Mass and Bond (1996). The along-coast boundary layer structure shows a tilt and

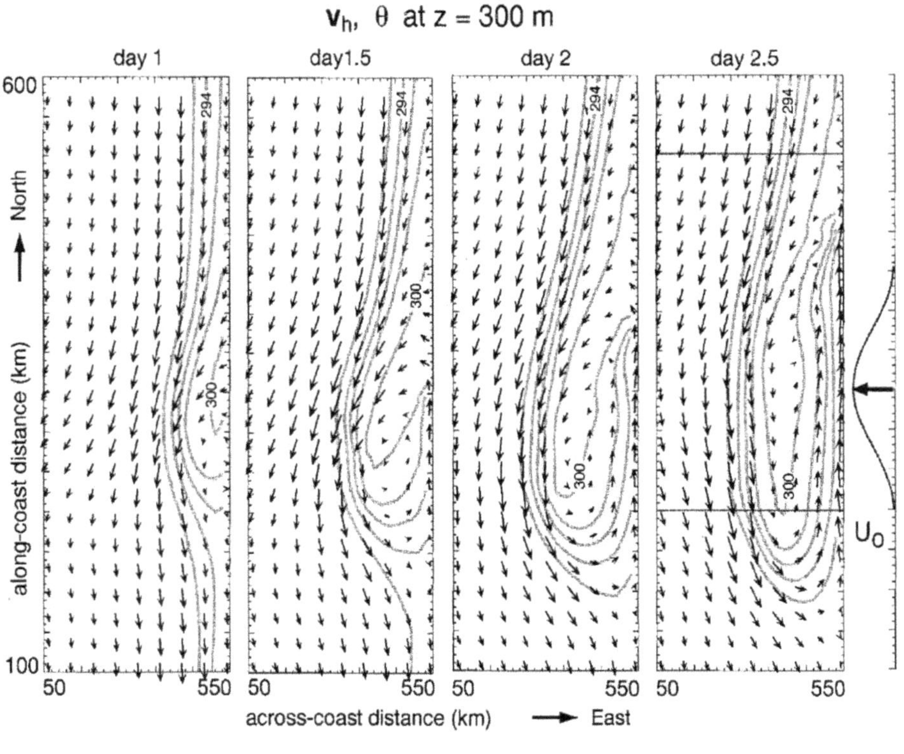

Fig. 7.12 Horizontal distribution of potential temperature and winds at 300 m in a numerical simulation along an idealized coastline, which was forced by a constant cross-coast flow illustrated on the right side of the figure. From Skamarock, Rotunno, and Klemp (1999). ©American Meteorological Society. Used with permission.

thickening of the inversion from south to north that supports a favorable pressure gradient to spawn a Kelvin wave with low inversion base height to the north.

This basic evolution is captured very well by Skamarock, Rotunno, and Klemp (1999) as shown in Fig. 7.12. The figure depicts the evolution of the boundary layer potential temperature (gives a similar picture to pressure) and the near-surface winds over a 36-hour period. The development of a coastal low is evident on day 1 which shows the separation of the strong northerly flow from the coast and its adjustment around this low center. To the south of this low the flow turns coastward where it is blocked by the topography. At the 1.5-day time period, a narrow zone of higher pressure and coastal southerly winds has developed, similar to that observed in the June 1994 case. This is the Kelvin wave being initiated. As the Kelvin wave

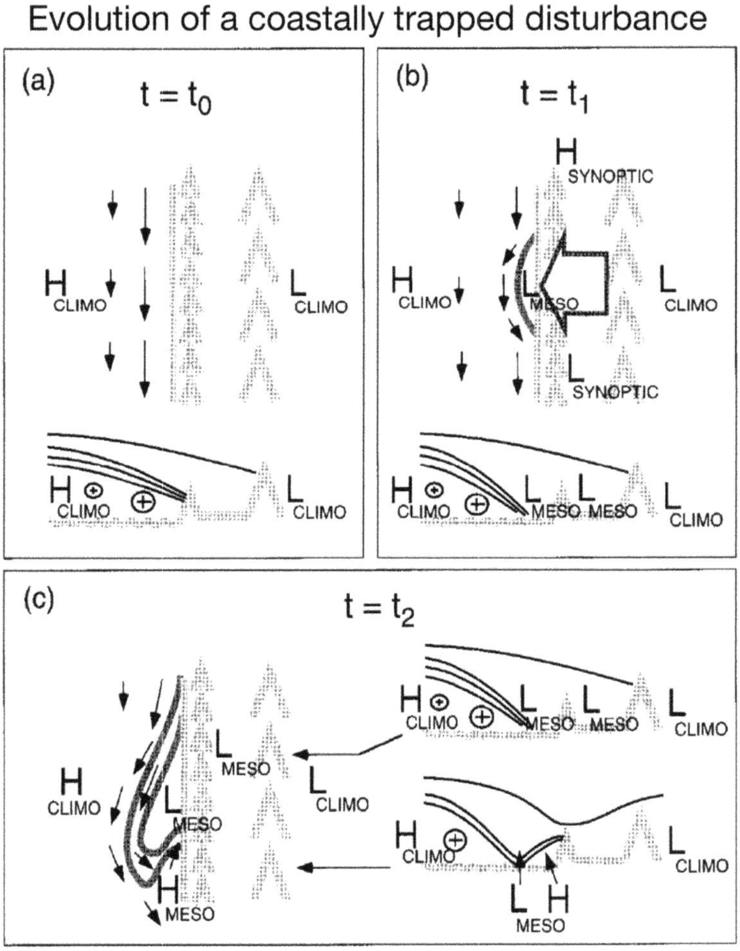

Fig. 7.13 Schematic summary of the evolution of a coastally trapped wind reversal forced by flow across the coastal topography. Evolution is shown as (a) initial time prior to offshore flow, (b) at time of initial offshore flow, and (c) after Kelvin wave has started. (From Skamarock, Rotunno, and Klemp (1999).) ©American Meteorological Society. Used with permission.

propagates north, the coastal low migrates offshore but stays at essentially the same location along the coast, which is consistent with the observations. A key result from this simple model is that the wind reversal initiates south of the strongest offshore flow and then propagates past and below the strong offshore flow (which occurs above the boundary layer) as it moves north. The profile of offshore flow is important to give the localized forcing needed to spawn the Kelvin wave.

This CTWR evolution is depicted schematically in Fig. 7.13 which illustrates the evolution from the climatological conditions to a coastal Kelvin wave due to the offshore flow. The cross coast cross-sections show how the boundary layer evolves to produce the upward tilt toward the coast characteristic of a Kelvin wave. While this structure is simplified compared to the structure in an actual CTWR, the interaction of synoptic scale and mesoscale response can be extended to capture more of the three-layer characteristics that were observed without any loss of understanding. This conceptual model is very simple but allows for some of the observed variation in events, including non-propagating events where the offshore flow does not spawn a Kelvin wave due to a lack of turning of the flow back toward the coast south of a localized low.

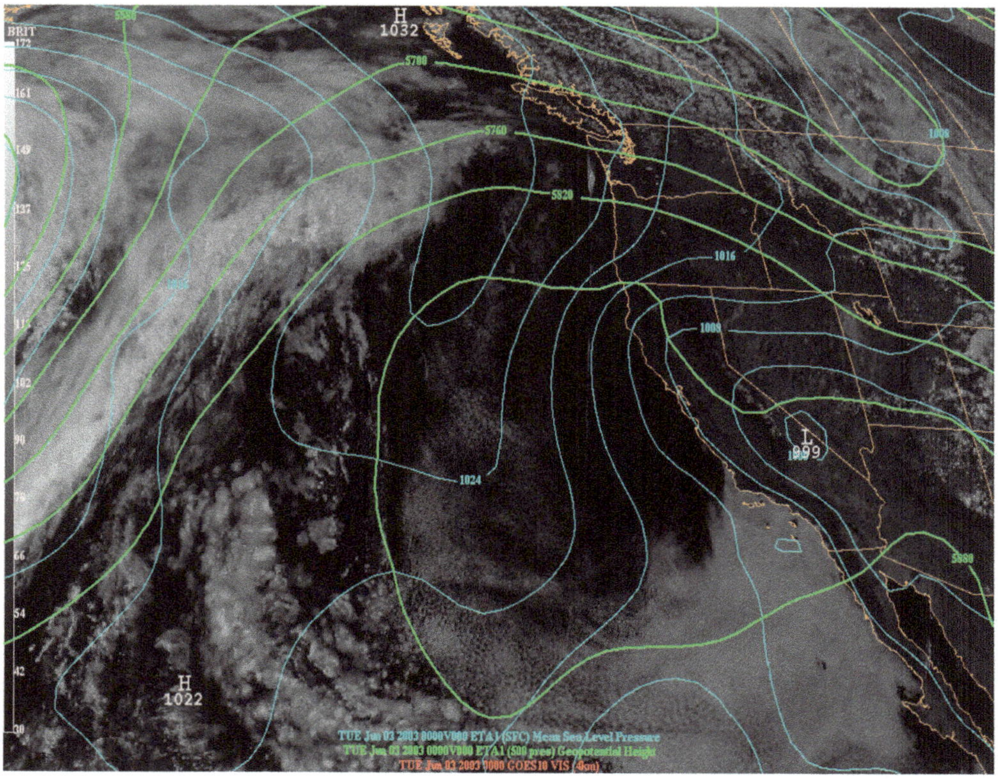

Fig. 7.14

Synoptic scale pattern and clouds prior to a CTWR along the US West coast. Coastal region is clear and occurs downstream from the 500 hPa ridge axis, which produces large-scale subsidence.

(a)

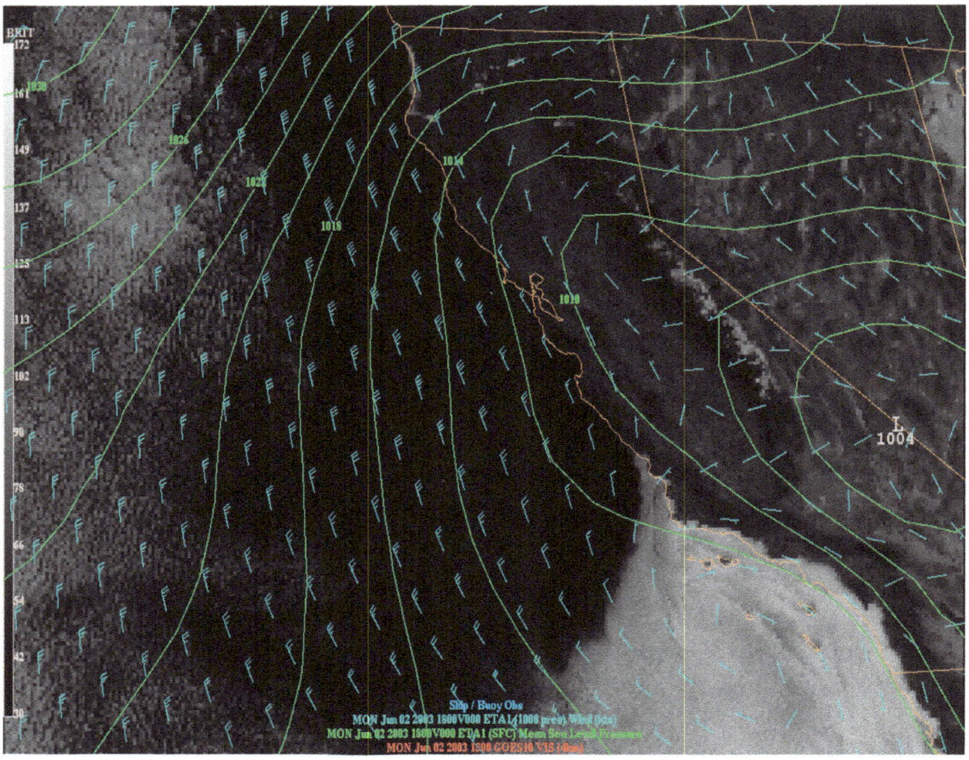

(b)

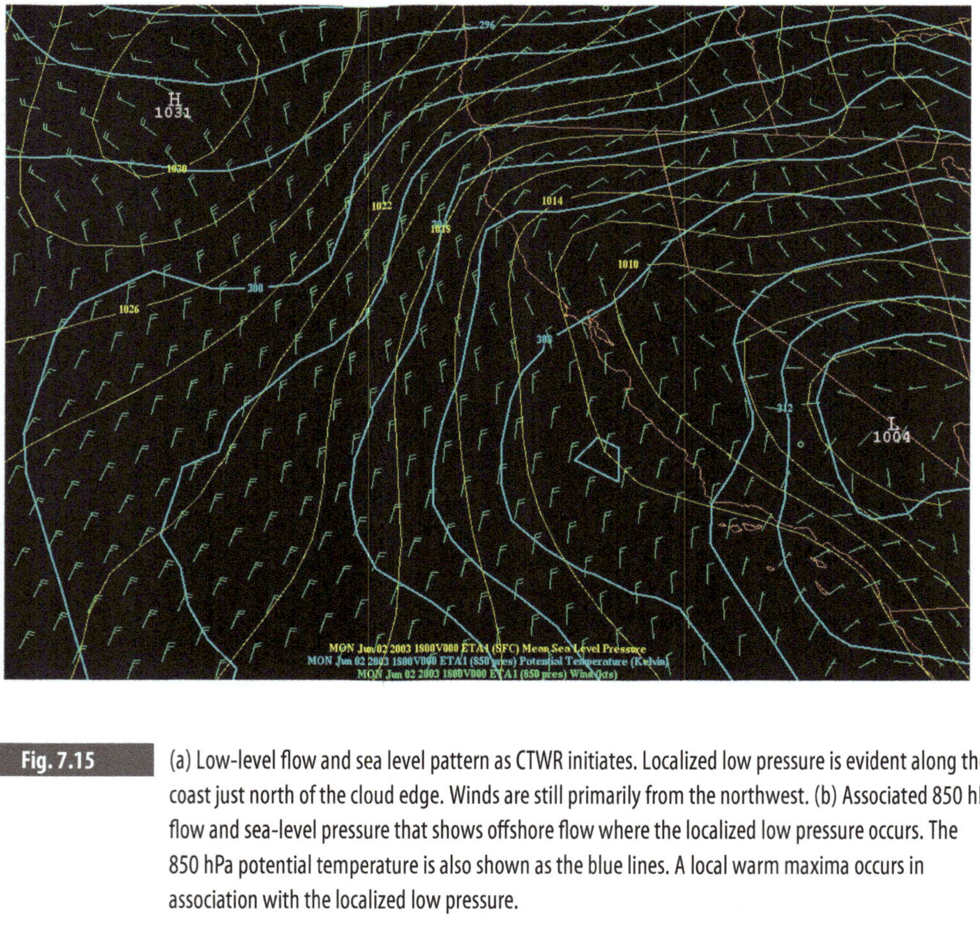

Fig. 7.15 (a) Low-level flow and sea level pattern as CTWR initiates. Localized low pressure is evident along the coast just north of the cloud edge. Winds are still primarily from the northwest. (b) Associated 850 hPa flow and sea-level pressure that shows offshore flow where the localized low pressure occurs. The 850 hPa potential temperature is also shown as the blue lines. A local warm maxima occurs in association with the localized low pressure.

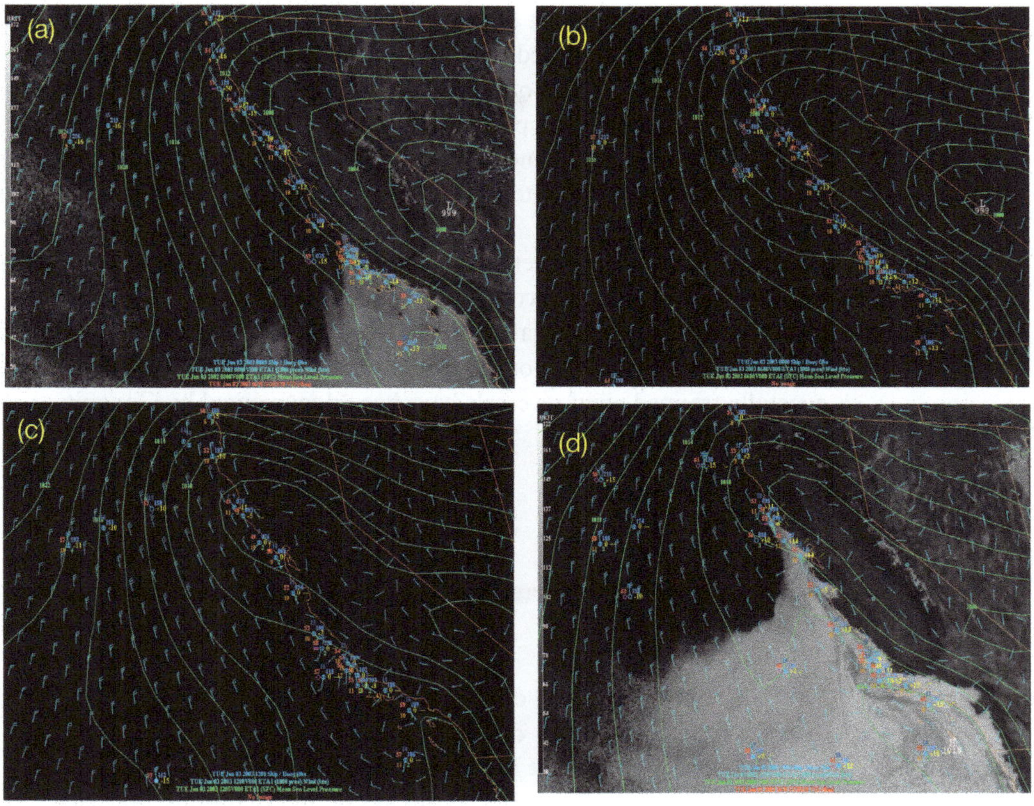

Fig. 7.16 Sequence of images showing CTWR moving north along the coast. Evolution of sea-level pressure and clouds from (a) time prior to offshore flow, (b) time near beginning of offshore flow, (c) time once CTWR has started, and (d) time after CTWR has propagated up the coast. The surface observations are also shown, and southerly winds occur in the cloud regions along the coast.

To illustrate the development and evolution of coastally trapped wind reversals, consider the following example. Prior to the event, the satellite image shows clear conditions along the US West Coast with upper-level ridging just offshore as seen in Fig. 7.14. The pressure gradient is largely perpendicular to the coast in California with no localized low pressure along the coast. The theory suggests that in order to initiate a wind reversal, there must be localized offshore flow that produces a localized region of low pressure along the coast. The sea-level pressure analysis and 850 hPa winds depicted in Fig. 7.15(b) show offshore flow over Northern California and across the central California coast, where a localized low-pressure center occurs. The Northern California flow does shift the pressure trough toward the coast. The central California flow results in warm potential temperatures offshore and a localized pressure minimum along the coast. This setup produces the onshore flow to the south of the localized low-pressure region as seen on the surface wind observations and satellite image in Fig. 7.15(a). Shortly after this time, the winds along the coast become southerly and low-level clouds form. The clouds and leading edge of the southerlies propagate up the coast as seen in the sequence of images in Fig. 7.16. Note that the coarse resolution CFSR fails to produce the coastal mesoscale ridge and does suggest that the local low moves north. High-resolution model simulations show a much more definite mesoscale pressure ridge, characteristic of a Kelvin wave.

While CTWRs have been most studied along the US West Coast, they have been observed to occur along other mountainous coastlines elsewhere. South America, Alaska North Coast, South Africa, and elsewhere have evidence of CTWRs occurring. Even along the north coast of Spain, coastally trapped wind reversals have been documented by Gangoiti et al. (2023).

7.5 Exercises

7.1 Calculate the nominal cross-coast length scale and propagation speed of a coastally trapped disturbance, assuming the depth of the disturbance is 300 m and the Brunt-Vaisala frequency is $0.04\ s^{-1}$.

7.2 Coastally trapped wind reversals occur in conditions that can produce downslope flow due to pronounced cross-coast(mountain) flow through a deeper layer. Explain why and what cross-coast, vertical profile of wind is mostly likely to produce the necessary coastal warming and localized low pressure to initiate the reversal (consider what you know about mountain waves). Explain whether this profile would be required if the coastal low were due to warm advection instead of downslope warming.

7.3 Calculate the along-coast pressure perturbation if the atmosphere warms by 2°C in the 800 hPa to surface layer. If this pressure perturbation occurs over a distance of 200 km (center to edge), calculate the geostrophic flow toward the coast. Assume $f = 10^{-4}\ s^{-1}$. Will this flow be blocked if the Brunt-Vaisala frequency is $0.04\ s^{-1}$ and the inversion height is 300 m?

7.4 Explain why stratus surges occur when offshore flow eliminates the marine layer inversion and pushes the subtropical high away from the coast.

8 Lee Troughs and Eddies

The development of lee troughs, windward ridges, and eddies around topographic features is an important aspect of flow over mountainous coastlines. As noted in Chapter 4, the flow interaction with topography can be divided into those characterized by low Froude numbers that are associated with flow blocking effects and those characterized by higher Froude numbers where the flow can more easily go up and over topographic features. The characteristics of high Froude number flows can be rather complex and depend strongly on details of the up- and downstream stratification as well as the Rossby number associated with the topographic barrier. For higher Rossby number flows, the flow does not adjust geostrophically to any induced perturbations and tends to produce features such as windward ridging and lee troughing in the sea-level pressure pattern. In this chapter, high Froude number flows around relatively isolated topography will be examined. Even when the Froude number is low and may support upstream blocking, the flow around more isolated topography can produce a variety of responses in the lee, which will also be examined in this chapter.

8.1 Description of Troughs and Eddies

Lee troughs and eddies are low-pressure regions on the lee side of topographic features. The low pressure forms due to the flow over the topography, producing warming as a consequence of downslope flow. This low-pressure center sets up a cyclonic circulation in the lee of the topography to produce a cyclonic eddy. Lee troughs and eddies have been observed along many coastlines and often owe their existence to specific geometries of the coastline and coastal topography. For example, the Catalina eddy in Southern California occurs in a region where the coast bends away from the background northwesterly flow. Oftentimes, bays along a coastline can form cyclonic eddies within them due to cyclonic shear of strong low-level flow. This has been observed in the Monterey Bay along the California coast.

8.2 Basic Mechanisms to Produce Lee Troughs and Eddies

The basic mechanism by which lee pressure troughs and the associated larger-scale eddies form is associated with the thermodynamic response to flow over a mountain. For a long mountain barrier where air does not appreciably flow around the sides of the mountain, the response depends strongly on the lee side stratification and the ability of the air to descend or be forced down in mountain waves on the lee side. This problem will not be considered in this chapter as the extended pressure trough does not typically result in cyclonic eddies in the lee of the topography. Instead, our focus will be on the flow around more isolated mountain features, where the flow can and does go around the mountain as well as over it. The finite scale to the cross-stream length of the mountain is important in producing the observed eddies and cyclonic circulations.

An excellent mathematical treatment of a simplified version of the interaction of a uniform flow with a single bell-shaped mountain can be found in Smith (1980) and Smith (1982). The full treatment of the problem can be found in those papers and will not be repeated here. A few key highlights from the Smith (1980) results are noted below and provide a simple basis for understanding the characteristics of flow over isolated mountain features. The results from Smith (1980) and Smith (1982) are based on a simplified set of linear perturbation equations using a uniform flow impinging on a bell-shaped mountain given by the function below.

$$h(x, y) = \frac{H}{\left(1 + \frac{r^2}{a^2}\right)^{3/2}}. \tag{8.2.1}$$

The parameter a represents the mountain half-width, $r^2 = x^2 + y^2$, and H is the maximum height of the mountain. For large Rossby number flows, Smith shows that the zeroth-order solution (leading terms in a Rossby number expansion) for the sea-level pressure and surface wind components are as follows:

$$p(x, y, 0) = -\rho NHU_0 \frac{x}{a}\left(1 + \frac{r^2}{a^2}\right)^{-3/2}, \tag{8.2.2a}$$

$$u(x, y, 0) = NH\frac{x}{a}\left(1 + \frac{r^2}{a^2}\right)^{-3/2}, \tag{8.2.2b}$$

$$v(x, y, 0) = NH\frac{y}{a}\left(1 + \frac{r^2}{a^2}\right)^{-3/2}. \tag{8.2.2c}$$

These solutions depend upon the incident flow speed U_0, Brunt-Vaisala frequency N and mountain height H plus a shape factor that is based upon

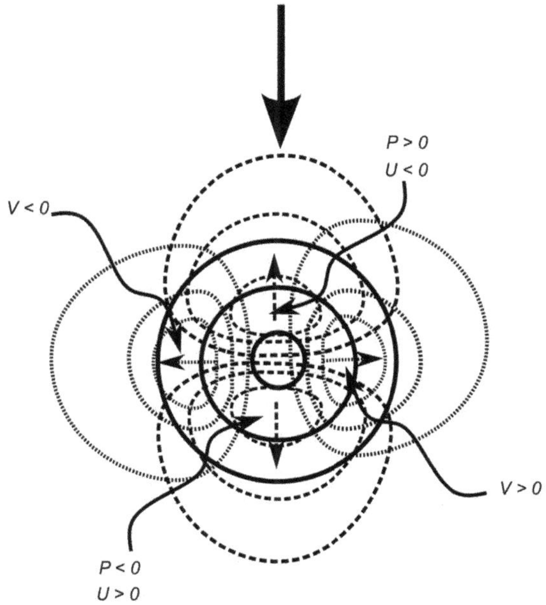

Fig. 8.1

Response of flow over a simple bell-shaped mountain (solid contours) showing the distributions of the perturbation pressure (dashed contours), U (dashed contours and arrows), and V (dotted contours and arrows) wind components. Incident flow is the heavy solid arrow. Adapted from Smith (1980).

the characteristic shape of the mountain. This basic solution is symmetric around the mountain and looks like that diagrammed in Fig. 8.1. The pressure response is shown by the dashed lines and produces high perturbation pressure on the upwind side and low perturbation pressure on the downwind side. The perturbation zonal wind has the same distribution but is opposite in sign, which produces opposing flow on the upwind side and enhancing flow on the downwind side. The perturbation meridional winds show flow away from the topography on both the left and right sides. This represents the tendency of the flow to go around the mountain.

This response shows a windward ridge and lee trough in the pressure response, deceleration on the upwind side and acceleration on the downwind side in the flow across the mountain (u), as well as flow deflection around the flanks of the mountain. The results provide a straightforward manner to consider the nature of the pressure response, which is composed of an amplitude factor (ρNHU_0) and a mountain shape factor $\left(\left(1+\frac{r^2}{a^2}\right)^{-\frac{3}{2}}\right)$. While this simple solution does not necessarily extend to arbitrary mountain shapes, it does suggest that the amplitude of any induced flow perturbations are given by the incident flow characteristics and the distribution of that response is likely to depend upon the shape of the topographic feature. This zeroth-order solution does not include any adjustment of the flow to the perturbations, which can be included by adding the next terms in the expansion of the solution.

Smith (1982) includes the first-order effects of Coriolis turning due to the perturbation pressure field. These first-order effects alter the basic solution above in the following manner.

- The pressure distribution is unchanged by the Coriolis effects.
- The cross-stream flow (v') deflection is positive everywhere (pointing to the left), and so this adds to the deflection on the left side of the mountain and decreases the deflection on the right side. The response is consistent with an ageostrophic acceleration toward the background low pressure.
- The cross mountain flow changes sign across the flow on either side of the mountain, which enhances the flow across the mountain on the left and weakens it on the right of the mountain.

The first-order solution impacts the wind field but not the pressure pattern. Again, the pressure response shows a windward ridge and lee trough as depicted in Fig. 8.2. A lee trough and windward ridge are clearly observed, and there is a strong cross-mountain pressure gradient. The flow deflects around the mountain to the left as well as going over the mountain. Presumably, the response to a more complex mountain shape can be made up of a combination of this response whose amplitude is given simply by ρNHU_0. The perturbation pressure response produces a cross-mountain pressure gradient along the axis of the incident flow. When this is added to the background pressure distribution, the total pressure pattern also shows a strong cross-mountain pressure gradient as the isobars are rotated to be perpendicular to the incident flow as shown in Fig. 8.2 for the simple solution. The degree to which the total pressure gradient aligns with the flow

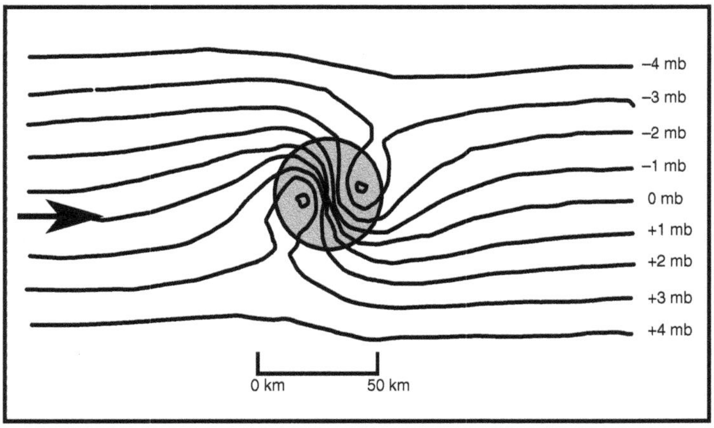

Fig. 8.2 Idealized total pressure response to flow over the bell-shaped mountain shown in Fig. 8.1 from Smith (1982). Background pressure gradient is uniform with low pressure at the top and high pressure at the bottom of the figure. This produces the incident flow from the left that induces the perturbation response with windward ridging upstream (left) and lee troughing (right). ©American Meteorological Society. Used with permission.

depends on the relative magnitude of the perturbation pressure gradient compared to the larger-scale background pressure gradient. Figure 8.3 (a) and (b) show real-world examples where the pressure gradient is largely across the mountain in the direction of the incident flow. In both examples, the pressure distribution over the topography results in a pronounced pressure gradient across the topography primarily along the incident flow axis, southerly for Iceland (Fig. 8.3 (a)) and northwesterly for New Zealand (Fig. 8.3(b)). These real-world examples show this characteristic pressure gradient across the mountain along the axis of the flow that signifies a topographic response.

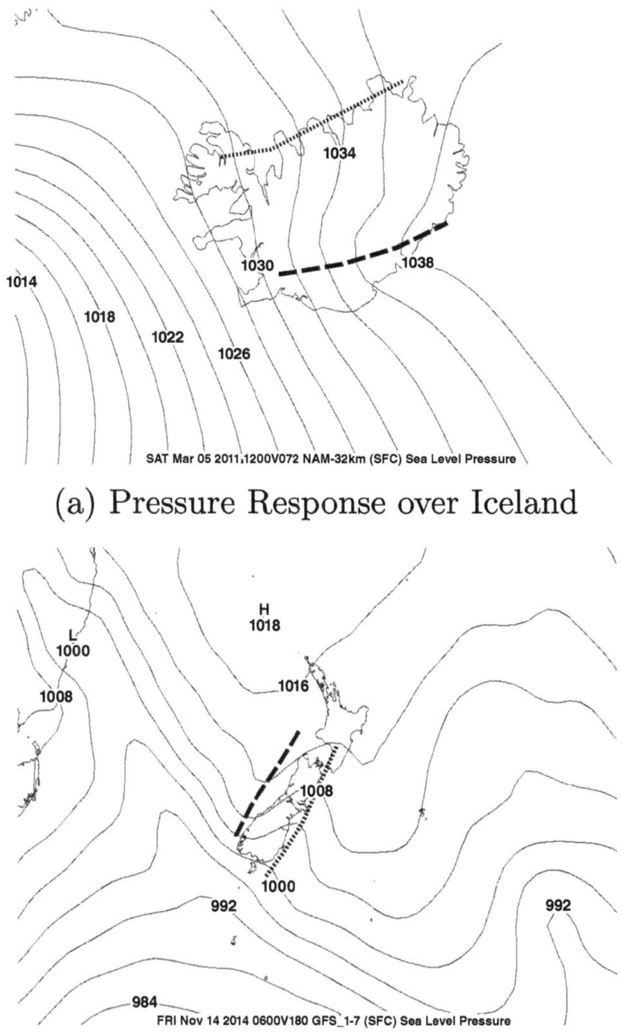

(a) Pressure Response over Iceland

(b) Pressure Response over New Zealand

Fig. 8.3 Observed total pressure distribution around (a) Iceland and (b) New Zealand during periods of flow across the topography. Windward ridge and lee trough are marked by the dashed and dotted lines, respectively

8.3 Flow around Isolated Topography

For isolated topographic barriers such as mountainous islands, the three-dimensional flow includes the flow that goes around the edges of the barrier as well as what may go over the top. Several common responses are observed when air flows over/around isolated topography. For the portion of the flow that is forced over the barrier, mountain waves can be excited as described in Chapter 4. The nature of the response in the lee depends on the wind and stratification characteristics of the incident flow and the downstream environment. If the boundary layer is capped by an inversion, which often occurs over many parts of the ocean region, then the height of the inversion relative to the height of the mountainous island determines the potential response. When the inversion occurs near mountain top, then flow over the island in this layer produces mountain waves that will be trapped as the stratification decreases sharply above the inversion. The waves propagate down stream from the island barrier to form a wake. The waves occur along the edge of the wake where the wind speed is favorable to maintain the trapped wave. In the immediate wake of the island, the wind speed tends to decrease, preventing the mountain wave from persisting. This is illustrated in the satellite image shown in Fig. 8.4. These island wake clouds and those that occur at the front edge of the island are analogous to the bow wave and wake produced by a ship in the ocean.

When the incident flow and stratification are such that flow over the mountain is prevented, such as when the inversion is below mountain top. In this case, horizontal shear instabilities can develop in the lee with strong flow along the island edge and weaker flow in the lee. Karmen vortices may occur downstream in the lee of the island as seen in the satellite image in Fig. 8.5. These lee vortices are produced if there is less flow over the topography and flow is forced around the barrier. The strength of the shear that develops along the flow in the lee of the barrier determines whether the flow becomes unstable to produce vortices. These vortices are shed by the barrier and propagate downstream to form a more extended stream of cyclonic and anticyclonic eddies in the lee.

8.4 Warm-Season Eddies

Although the dynamics of lee troughs and eddies do not depend on the season, the synoptic patterns that produce certain eddies are often determined seasonally. One example of a summer-time eddy that may be related to flow response to coastal mountains is the Catalina Eddy that develops

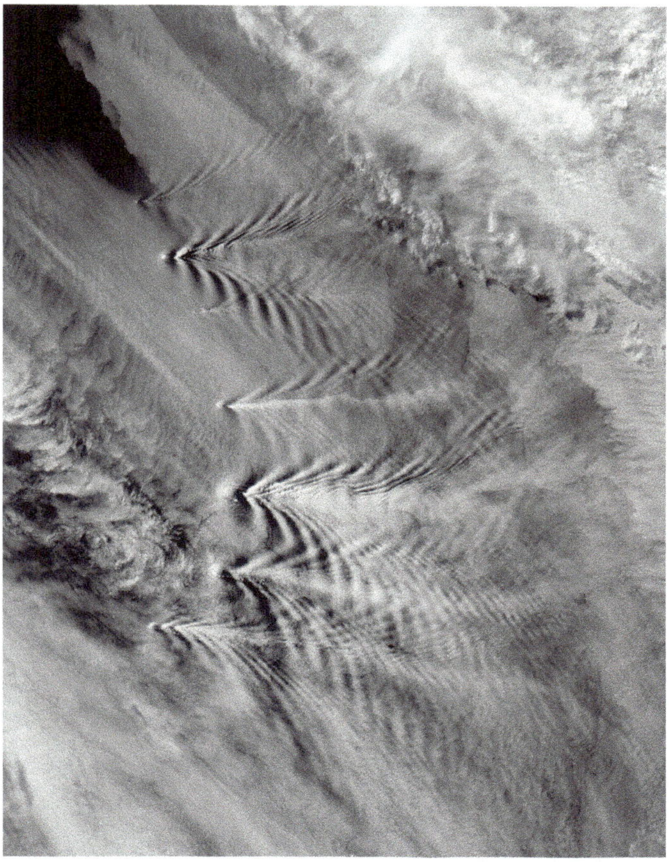

Island wake clouds around the South Sandwich Islands. Flow is from the west, resulting in trapped mountain waves downstream from the islands.

Fig. 8.4

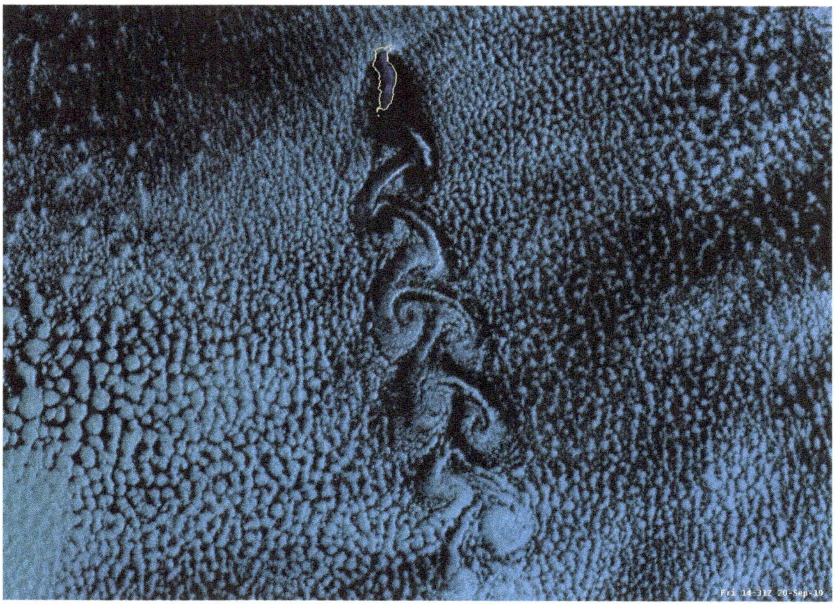

Karmen vortices occurring to the south of Guadalupe Island off the coast of Mexico. Flow from the north and blocked by the mountain island resulting in shear vortices in the flow around the island.

Fig. 8.5

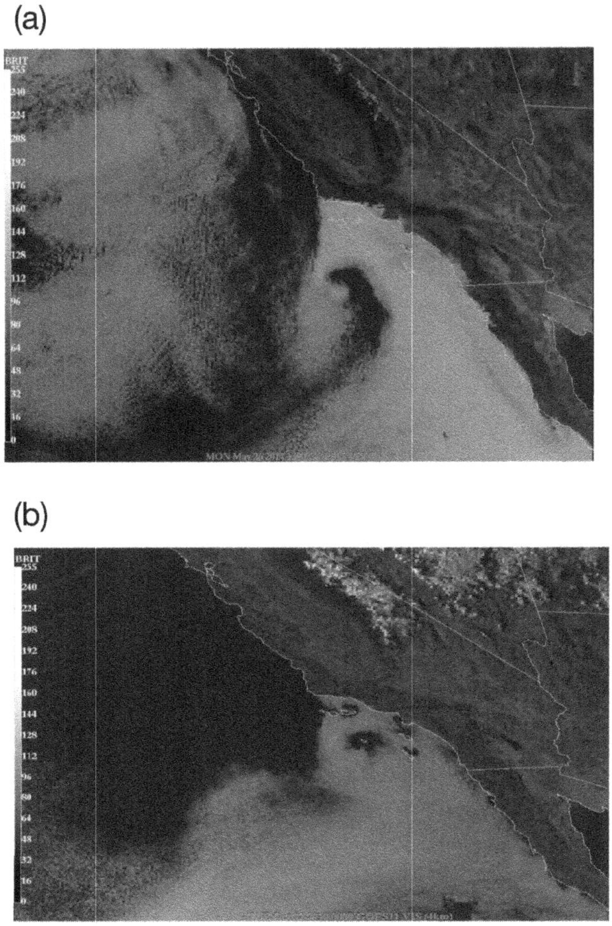

Fig. 8.6 Satellite images of several Catalina Eddy events in the Southern California Bight region. (a) Well-defined cyclonic cloud pattern from May 26, 2014 at 1500 UTC. (b) A more filled-in cloud pattern from October 12, 2010, at 0000 UTC.

in the Bight region of the Southern California coast. This feature has several aspects that go beyond the topographic response, but the initiating mechanism for the development of these features seems to be related to orographic forcing in most cases. The eddy is best characterized by the cyclonic circulation evident in the marine stratus that occurs over the region during much of the summer months. Figure 8.6 shows several examples of Catalina eddies as depicted by the stratus field that shows a cyclonic circulation. The overall tendency for cyclonic circulation in the region is suggested by the composites of station observations in the Southern California region based on 50 events from Mass and Albright (1989), which is shown in Fig. 8.7. The coastal winds show a strong diurnal sea breeze signal, but during the nighttime periods when the sea breeze is weak, a cyclonic circulation is suggested. This eddy composite is in contrast to a

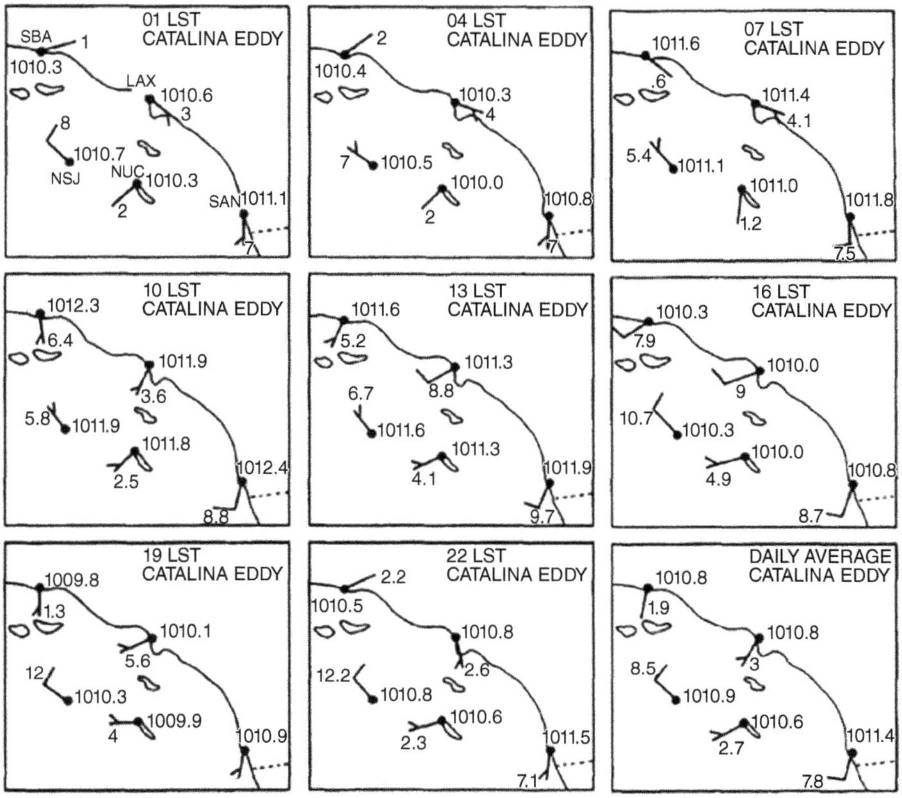

Composite surface wind and sea-level pressure for 50 Catalina Eddy events from Mass and Albright (1989). ©American Meteorological Society. Used with permission. **Fig. 8.7**

similar composite of noneddy days, where the offshore observations are similar, but the coastal sites fail to develop a southerly flow to create the cyclonic circulation. While the sea breeze signature is equally strong, the nighttime hours fail to develop much of any cyclonic circulation. The coastal winds near San Diego seem to be most crucial for defining an eddy in these composites. If the winds are southerly, then an eddy seems to be present, whereas westerly winds are associated with noneddy days. The eddy itself is associated with low pressure over the center of the Bight region after the eddy has matured. This position does not fit the idea of flow interacting with the coastal mountains to force an eddy so far from the coastal topography. To better understand the possible role of the flow interaction with topography in Catalina eddies, Mass and Albright (1989) examined the evolution of a single event.

The development of the Catalina eddy described by Mass and Albright (1989) is similar to the development of coastally trapped wind reversals described in Chapter 7. In the case of the Catalina eddy, the flow around 850 hPa near mountain top turns northerly or slightly northeasterly across the mountains north of the Bight region. This cross-mountain flow sets up downslope warming and a lee trough in the northern Bight region. As a consequence of the lee troughing, the pressure gradient reverses along the Southern California coast from San Diego to Santa Barbara. This pressure gradient reversal initiates a southerly flow along the coast due to a blocked response in the cross-coast flow. The southerly flow along the coast and northwesterly flow offshore to the west set up a larger-scale cyclonic circulation within the Bight region. This is a Catalina eddy. The continued northerly flow above the marine boundary layer advects warm air southward into the center of the Bight region to further strengthen the low-pressure center and cyclonic circulation. Skamarock, Rotunno, and Klemp (2002) modeled this pattern, similar to that done for coastally trapped wind reversals. They show lee troughing over the mountains to the north and the development of coastally trapped southerly flow along the southern California coast. This set up a larger-scale Catalina eddy circulation covering the entire Bight region. While this model describes many Catalina eddy events, observations suggest that mostly any flow pattern that results in warming over the center of the Bight region and an associated lowering of the pressure can set up the cyclonic circulation.

The following case illustrates the development of a Catalina eddy as the flow over the Southern California region evolves. The synoptic-scale evolution for this event that occurred on October 11, 2010, is depicted in Figs. 8.8–8.10. The 500 hPa evolution (Fig. 8.8) shows the approach and subsequent decay of an upper-level trough across the West Coast north of the Southern California region. There is no remarkable signature at 500 hPa that signals the initiation of a Catalina eddy on October 11. The upper-level trough is off the coast with a ridge axis extending across the central to southern California region on October 10. A surface cyclone occurs with the upper-level trough just off the British Columbia coast. As the trough moves east into Canada and the Pacific Northwest, it weakens and the upper-level ridge shifts eastward as well. This sets up zonal flow across the west coast that is very weak over the southern California region. At the surface, a weak inverted pressure trough occurs along the southern California coast on October 10, which strengthens on October 11 as the upper-level ridge shifts east. By 1200 UTC on October 11, a closed low-pressure center occurs in the southern California bight region, signifying the development of the Catalina eddy.

Crucial to the development of lee troughing to initiate eddy formation is the development of cross-mountain flow over the northern bight region. The 850 hPa evolution is depicted in Fig. 8.9 and shows that the subtropical high in the lower levels tends to amplify and move closer to the coast as the eddy develops. As the subtropical high amplifies, northerly flow develops and strengthens over the central California region to give cross-mountain flow in the northern bight. The Vandenburg sounding, just north of the region, at 1200 UTC October 11 indicates that northerly wind

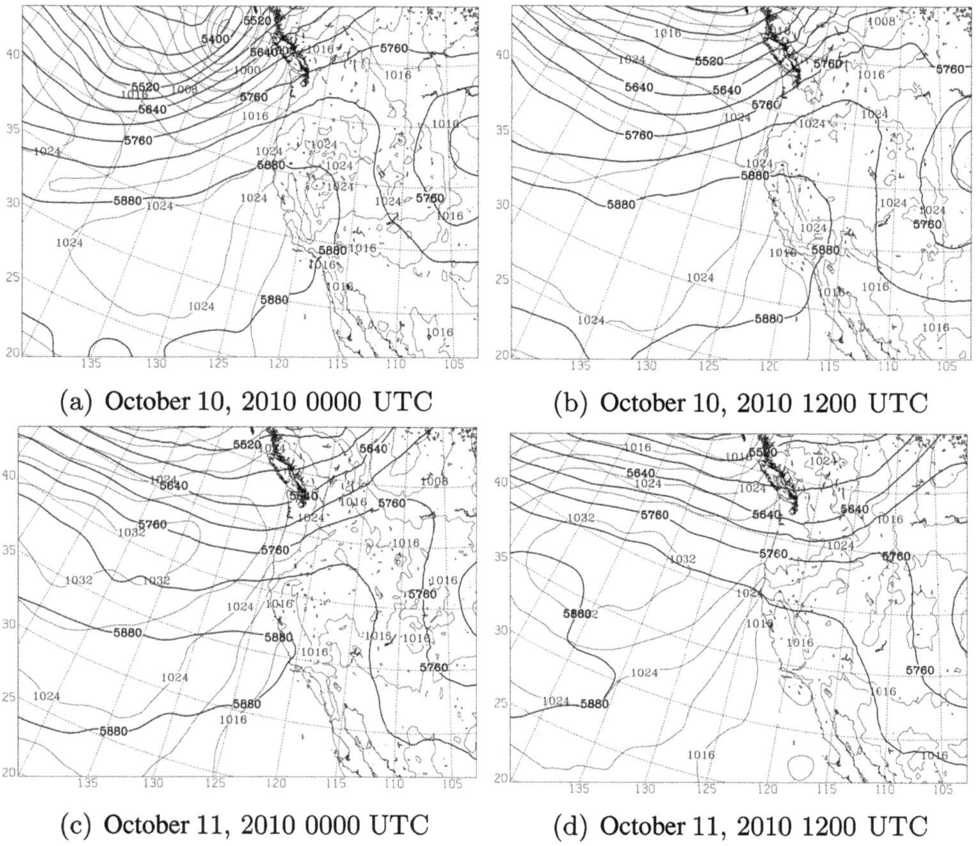

(a) October 10, 2010 0000 UTC

(b) October 10, 2010 1200 UTC

(c) October 11, 2010 0000 UTC

(d) October 11, 2010 1200 UTC

Fig. 8.8 Sequence of charts showing 500 hPa geopotential height (thick solid contours) and sea-level pressure (thin solid contours) from October 10 to 11, 2010 for a Catalina Eddy event.

has developed as the eddy becomes established. The sounding indicates a cross-mountain flow of 30 kts. The perturbation pressure response for this magnitude of cross-mountain flow is about +/− 3 hPa based on the linear theory of Smith. The observed perturbation pressure response, shown in Fig. 8.10, illustrates the lee trough over the northern bight and reversed pressure gradient along the coast to the south.

Mass and Albright (1989) have indicated that the development of the coastal southerlies that spawn the mesoscale cyclonic eddy once lee troughing is established represents a trapped response similar to the stratus surge we examined in the last chapter. A modeling study by Skamarock, Rotunno, and Klemp (2002) seems to support the characterization of the reversed flow along the coast as a trapped response that establishes the cyclonic circulation of the eddy. Their simulations did include a continuous coastal rise that allowed for a trapped response. The problem with this characterization is that the actual coast is not continuous mountains between San Diego and Santa Barbara, and so a trapped response would

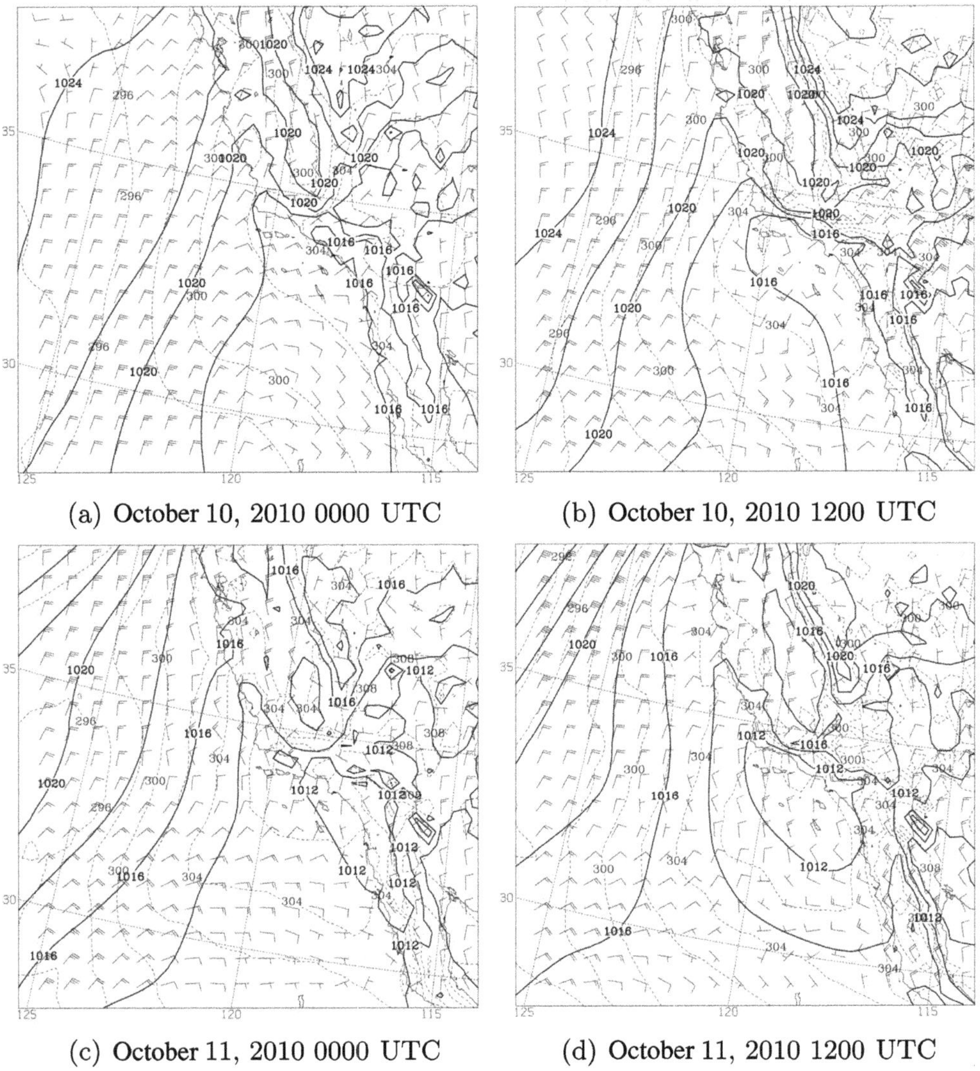

(a) October 10, 2010 0000 UTC (b) October 10, 2010 1200 UTC

(c) October 11, 2010 0000 UTC (d) October 11, 2010 1200 UTC

Fig. 8.9 Sequence of charts showing 850 hPa winds (barbs), sea-level pressure (solid contours), and 925 hPa potential temperature (dashed contours) from a October 10 to 11, 2010 Catalina Eddy event.

be hard to realize, as the flow can easily flow inland. The reversed pressure gradient is however sufficient to force a southerly component, particularly as the sea breeze decays even if it is not trapped.

The results from Skamarock, Rotunno, and Klemp (2002) provide a more complete characterization of the development of the cyclonic circulation. Their results using a variety of different cross-coast flow configurations, stratifications, and coastal topography all resulted in a cyclonic eddy. Crucial to the eddy formation was the establishment of a low-pressure region offshore over the bight region. Warming by the

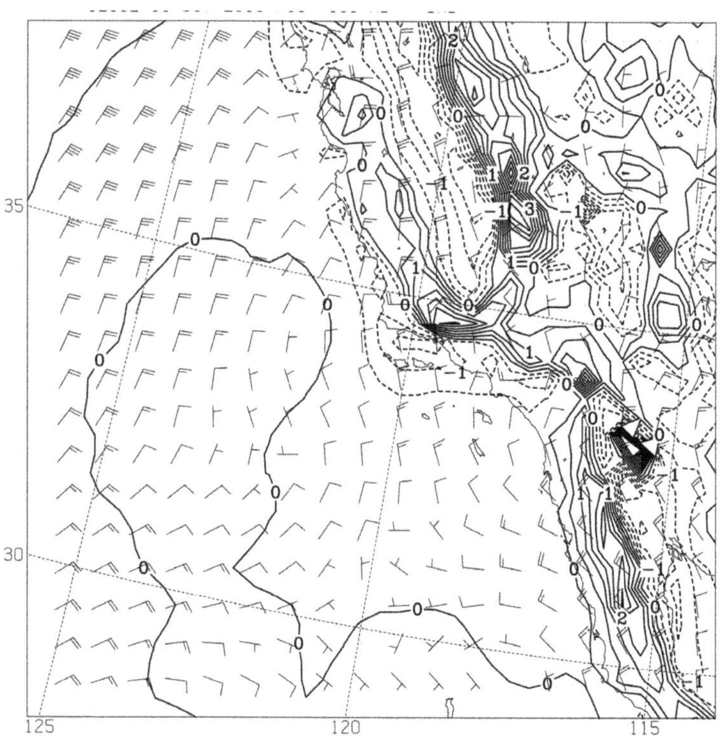

Fig. 8.10

Sea-level pressure perturbation (solid/dashed contours) and 850 hPa winds (barbs) at 1200 UTC October 11, 2010, as the Catalina Eddy initiates.

downslope flow and offshore advection of the warm air create the low pressure region. Given its persistent nature, the wind simply adjusts to the pressure field to create the cyclonic circulation. Other studies have also shown that the pressure is lowered offshore in the Bight as surface divergence is established due to the sea breeze and strong northerlies further offshore. This surface divergence lowers the inversion height over the center of the bight region to produce a low-pressure center. Over time, the winds respond to this pressure pattern in a quasi-geostrophic manner to establish a circulation around the low. This interpretation is consistent with observations that show the lowest inversion base height occurs offshore in the low center. The inversion drop is consistent with both surface divergence set up by the geometry and warm air created by the downslope flow and its advection out over the bight. The potential temperature at 925 hPa shown in Fig. 8.9 shows a warm region over the center of the bight consistent with a suppressed marine layer and subsidence over the area. The mechanisms that seem to strengthen the cyclonic circulation into a mature eddy seem to vary, but in many cases, the eddy is initially forced by the interaction of the flow across the coastal topography, particularly the topography north of Santa Barbara.

8.5 Cold-Season Eddies

The interaction of flows during the cool season also produce notable examples of windward ridging and lee troughing responses that produce lee eddies or vortices as well. Flow across the Olympic Mountains in the Pacific Northwest provides a clear example of this type of interaction with relatively isolated topography. Figure 8.11 depicts the topography in the Puget Sound region of western Washington and shows the relatively isolated character of the Olympic mountains, which may demonstrate a flow response somewhat similar to that described by the Smith (1982) theory for a bell-shaped mountain.

The mean pressure distribution for the period from April through June depicts the tendency of the region to experience sea/land breeze-type diurnal forcing. The land areas tend to show decreased pressure during the daytime period and higher pressure at night. This supports the weak onshore flow during the day and divergent offshore flow during the night over the Olympic Mountains. The magnitude of this circulation is small and does not substantially alter the larger impacts of flow interaction with the topography.

There are two prevalent flow patterns associated with the topographic interaction during the winter months for this region. One is that associated with events referred to as Hood Canal wind events. This type of event was named for a significant event that occurred in 1979, in which a floating

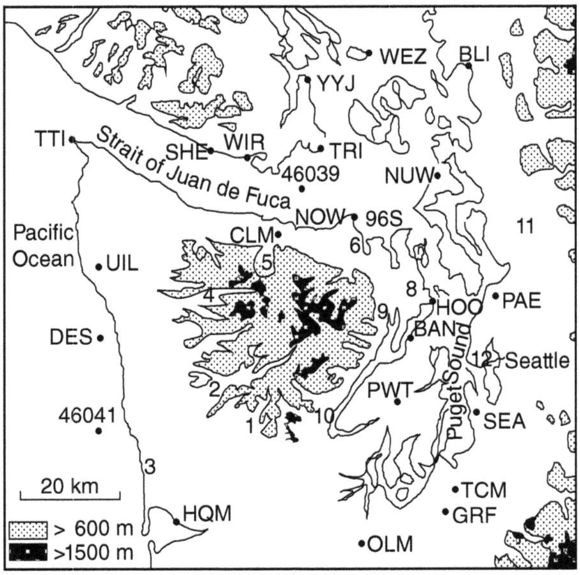

Fig. 8.11 Topography and surface stations around the Olympic Peninsula of Washington State from Ferber and Mass (1990). ©American Meteorological Society. Used with permission.

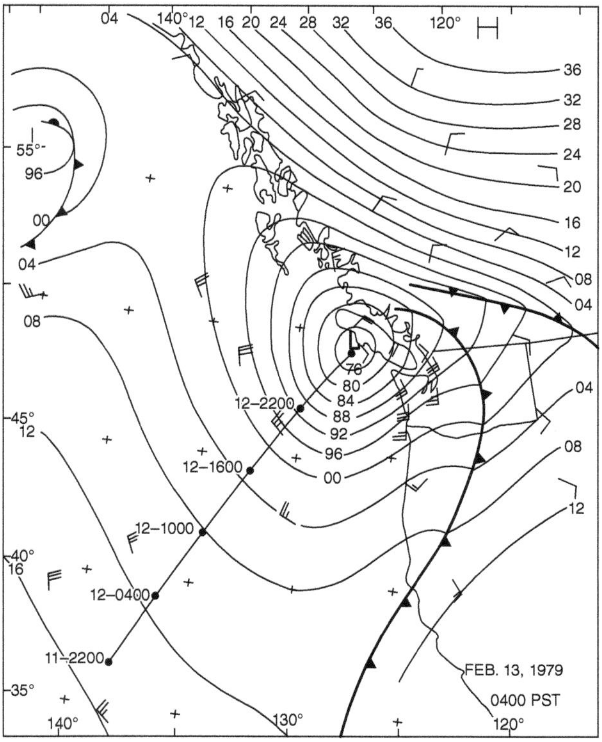

Fig. 8.12

Synoptic scale pressure analysis for the Hood Canal windstorm of February 1979. Strong low-pressure system occurs just west of Vancouver Island at 1200 UTC 13 February 1979. From Reed (1980). ©American Meteorological Society. Used with permission.

bridge at the mouth of Hood Canal, on the southeast side of the mountains, sank as a result of very strong winds and the associated high waves. This case is depicted in Fig. 8.12, which shows a strong synoptic-scale low-pressure system approaching the coast of Vancouver Island. The Salem Oregon sounding (not shown), to the south of the region, indicates a strong southwesterly flow in a nearly moist neutral environment, which is favorable for flow over any topographic features. A corresponding mesoscale pressure analysis in the Puget Sound region is shown in Fig. 8.13 and indicates an intense mesoscale low-pressure center on the east side of the mountains. The cause of this low-pressure region is lee troughing associated with the strong southwesterly flow across the Olympic Mountains. The induced strong pressure gradient resulted in very extreme down-gradient flow (in excess of 70 kts) that resulted in sinking the bridge. The bridge sinking was the result of the strong winds acting on the ocean in the narrow channel of the Hood Canal. This resulted in excess water pressure on the bridge as well as the strong winds on the structure above the water.

A more thorough analysis of similar type events done by Ferber and Mass (1990) demonstrates the characteristics of this type of flow interaction with the relatively isolated Olympic Mountains. A composite sea-level

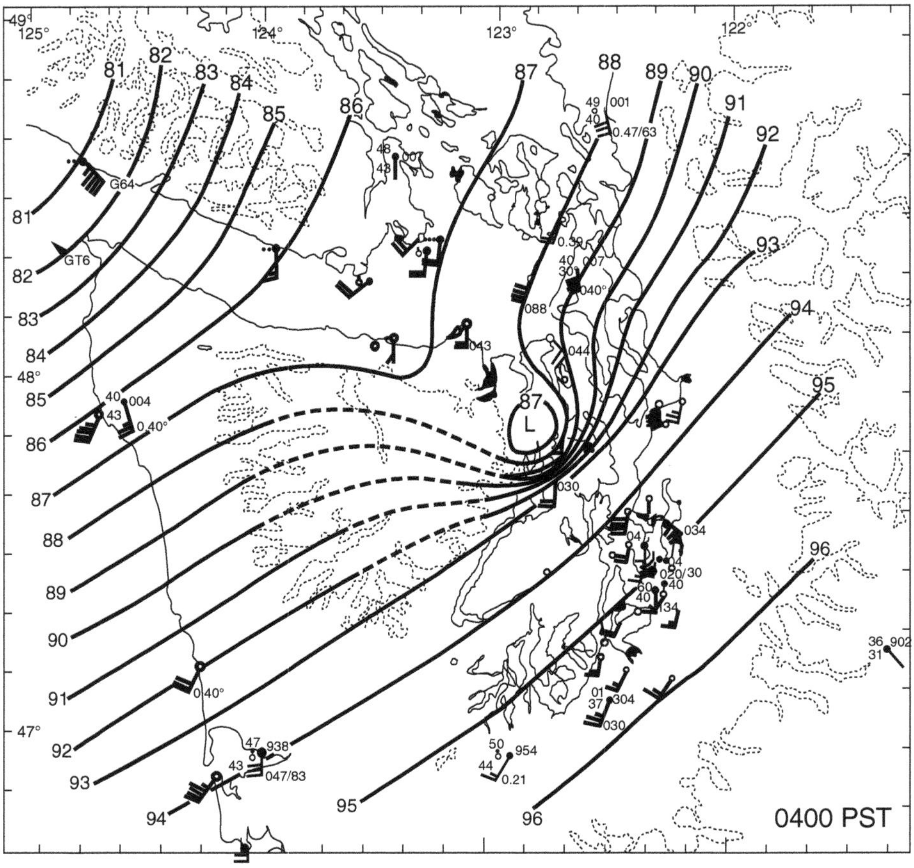

Fig. 8.13 Mesoscale pressure analysis for the Hood Canal windstorm of February 1979. A strong lee trough is seen to the northeast of the Olympic Mountains at 1200 UTC February 13, 1979, which produces a very strong pressure gradient to the southeast of the mountains. From Reed (1980). ©American Meteorological Society. Used with permission.

pressure analysis and the perturbation pressure are shown in Fig. 8.14. The analysis shows a characteristic upwind ridge and downwind trough both 6 hours prior and 6 hours after the peak amplitude to the event. The synoptic-scale conditions produce southwesterly flow across the mountains in a weakly stratified environment as suggested by the composite sounding from Quillayute shown in Fig. 8.15. The pressure response is similar to that implied by the linear theory of Smith (1982), although the windward ridge and lee trough response are not absolutely symmetric as suggested by the Smith theory. The asymmetry may be an artifact of the asymmetry in the mountain slopes on the west and east sides of the Olympic Mountains. The Smith theory does indicate a shape factor associated with the mountain is also important. An example of the synoptic-scale flow pattern and mesoscale perturbation pressure response (Fig. 8.16) are shown for a Hood Canal wind event. The strongest lee trough response is associated with 45–50 kt winds at 850 hPa wind as the synoptic-scale

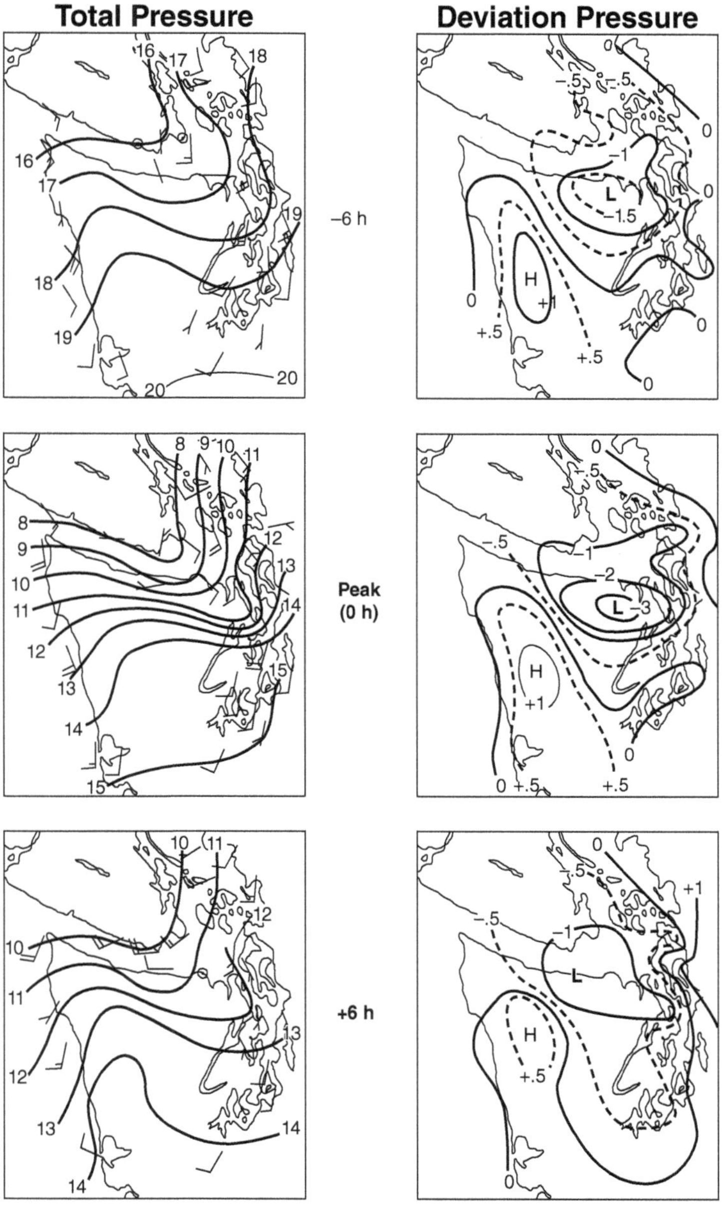

Fig. 8.14 Sea-level pressure (total and deviation) and surface wind composites for nine Hood Canal windstorm events from Ferber and Mass (1990). ©American Meteorological Society. Used with permission.

flow became strong and more directly out of the southwest than at earlier or later times in this case. Using the Smith (1982) formula for this case, the predicted pressure response is 3.6 hPa = ρNHU_0 assuming $H = 1000$ m, $N = 1.4^{-2}$, $U_o = 20$ m/s and $\rho = 1.3\,\text{kg/m}^3$. While this exceeds the observed perturbation response of about 2 hPa, it is consistent with the results of Ferber and Mass (1990) that found the linear theory to over-estimate the response by about a factor of 2.

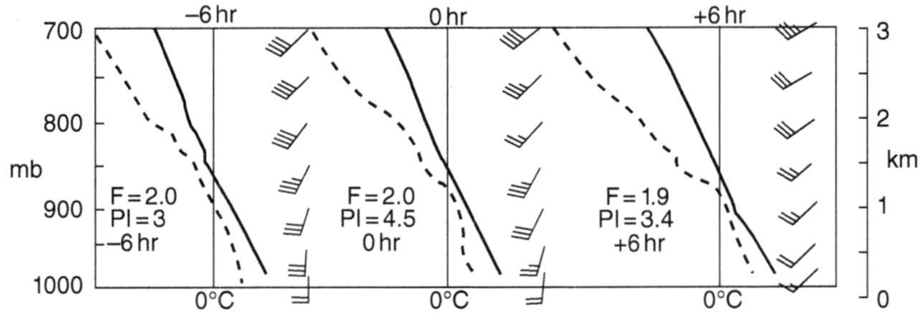

Fig. 8.15 Composite soundings from Quillayute, Washington for nine Hood Canal windstorm events from Ferber and Mass (1990). ©American Meteorological Society. Used with permission.

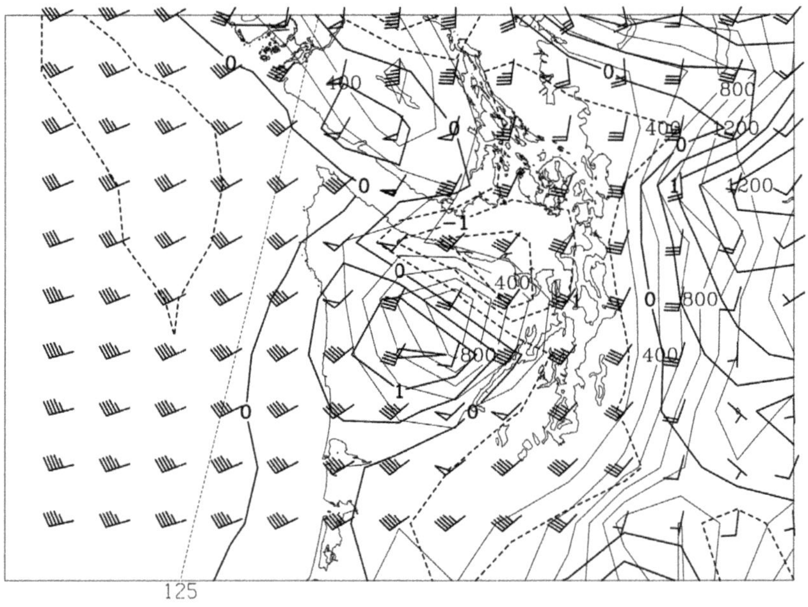

Fig. 8.16 Sea-level pressure perturbation (solid/dashed contours every 0.5 hPa) and 850 hPa winds (barbs) near the Olympic mountains (thin solid lines) for October 2013. The maximum positive and negative perturbations align with the southwesterly cross-mountain flow.

Another type of flow response occurs when the synoptic scale flow shifts to a more west to northwesterly direction. This type of flow is typically associated with a cold post-frontal flow and spawns a phenomenon referred to as the Puget Sound convergence zone. A composite of Puget Sound convergence zone events is shown in Fig. 8.17 and illustrates the tendency of the flow to diverge around the Olympic Mountains on the upwind side and then converge on the lee side in the Puget Sound region. The convergence is aided by the occurrence of a mesoscale lee trough in the region that is spawned by the northwesterly flow over the mountains.

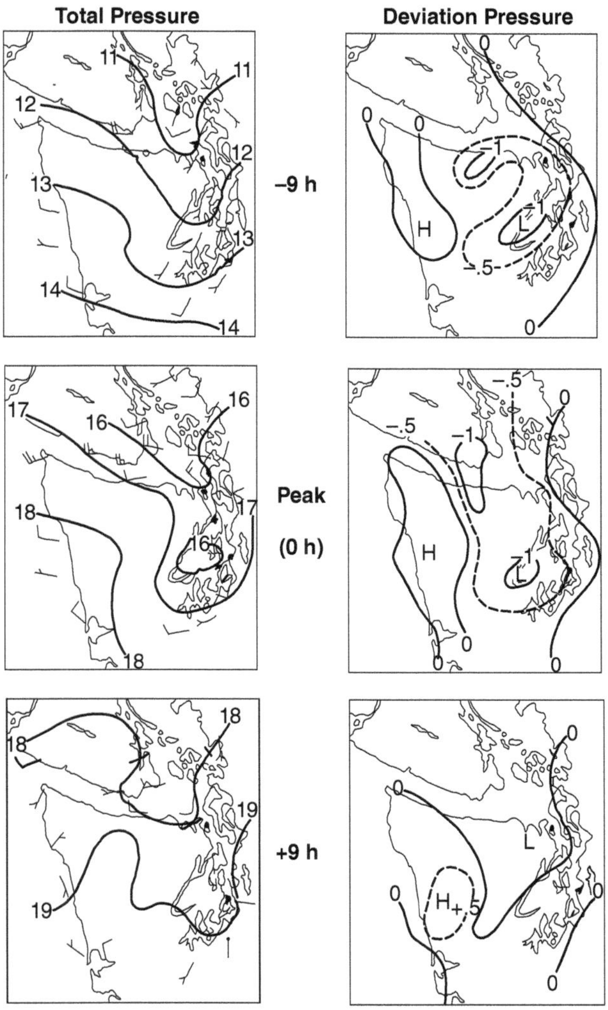

Fig. 8.17 Composite sea-level pressure (total and deviation) and surface wind for Puget Sound Convergence Zone events from Ferber and Mass (1990). ©American Meteorological Society. Used with permission.

The magnitude of the response is less than that seen for Hood Canal wind events due to the generally weaker synoptic-scale flow speeds. This is evident in the composite Quillayute soundings for these types of events shown in Fig. 8.18. Winds are generally less than 20 kts, and the static stability is typically moist neutral to above 800 hPa which supports flow over the mountains. A particular case is illustrated in Fig. 8.19, which shows the weak magnitude of the topographic response and almost nonexistence of any windward ridging effect in these events. The distribution of the lee trough shows a tendency to occur in a similar region to Hood Canal events, even though the flow is from a very different direction. This suggests an asymmetry in the mountain-induced response that goes beyond the Smith theory. The mountain shape seems to favor a response to the northeast as long as there is some component of the flow across the mountains in that direction. The lee troughing in the central sound is crucial for the southwesterly flow to force the convergence zone, and it only gets forced when the flow is strongest from the northwest.

The directional dependence of the flow response to the Olympic Mountains has been explored by Ferber and Mass (1990), and it fits the basic idea from the Smith (1982) theory. The flow response to the Olympic Mountains over a range of wind speeds and flow directions shows larger perturbations with higher wind speed and perturbations that align with the incident flow direction. As predicted from the Smith (1982) theory, as the wind speed increases the magnitude of the topographic response also increases. The figures however also show that the asymmetry of the Olympics play an important role in the location of the response by producing a lee trough on the northeast side through many wind directions. This dependence on the mountain shape is important and alters the magnitude of the response as well as shifting its location relative to the incoming flow direction.

The results from the interaction of the flow around the Olympic mountains suggest that a simple application of the Smith (1982) theory can provide insight into the first-order response to topography for high Froude number flow regimes. The development of windward ridges and lee troughs with a magnitude that depends on the wind speed and lies roughly along the axis of the incident flow is consistent with this theory. Ferber and Mass (1990) show that the magnitude of the response is most strongly dependent on the wind speed and less dependent on the stratification as long as the stratification supports flow over the topography. Their results also show that the actual response is typically about half that predicted by a simple application of the Smith (1982) theory. Presumably, nonlinear effects and stronger dependencies on the mountain shape account for these differences. However, a simple application of the basic response model using synoptic-scale conditions can provide an important check on mesoscale model forecasts or provide some mesoscale information when no other mesoscale forecasts are available.

While the results presented here apply to events that occur around the Olympic Mountains of Western Washington, the flow interaction with relatively isolated topography elsewhere produces some similar type events. Flow around larger islands, such as the Big Island of Hawaii also produces lee troughs and eddies. An important aspect of flows around islands like Hawaii is the degree to which flow goes over the island/mountain feature as opposed to simply being split around it. The flows over the Olympic Mountains described above results from a significant portion of the flow going over the isolated topography. This occurs when the stratification is sufficiently weak and the incident flow is strong. For islands like Hawaii, the mountains are tall and extend well above the capping inversion. This prevents significant flow over the mountains and forces the flow to go around the barrier. The result is the formation of both a cyclonic and

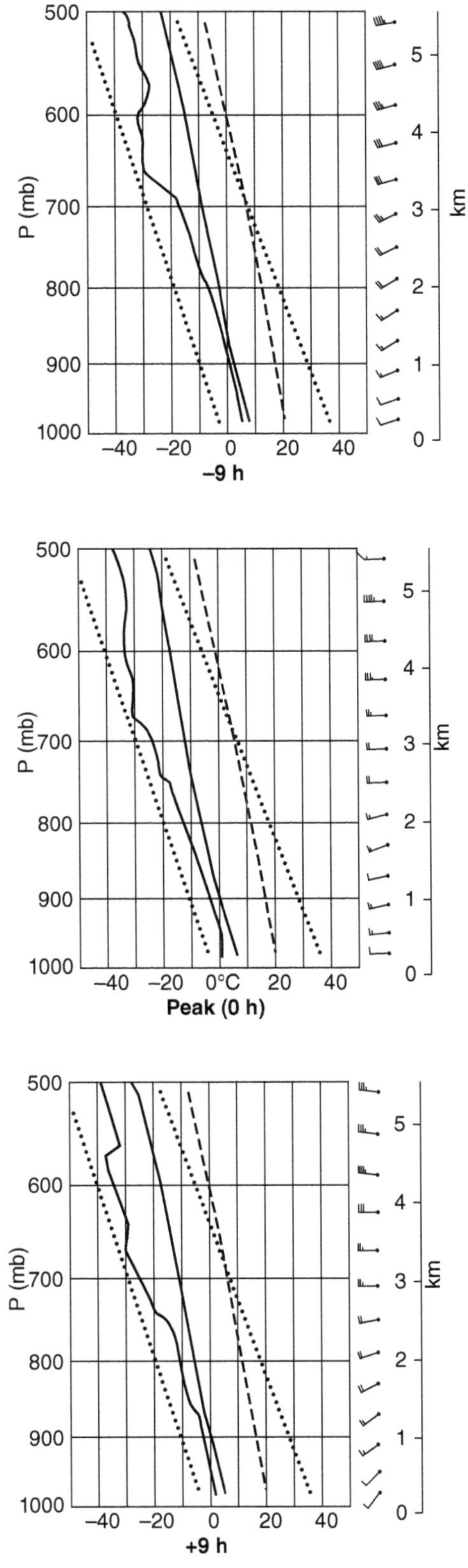

Composite Quillayute, Washington soundings for Puget Sound Convergence Zone events from Ferber and Mass (1990). Dotted lines are dry adiabats and dashed lines are moist adiabats. ©American Meteorological Society. Used with permission.

Fig. 8.18

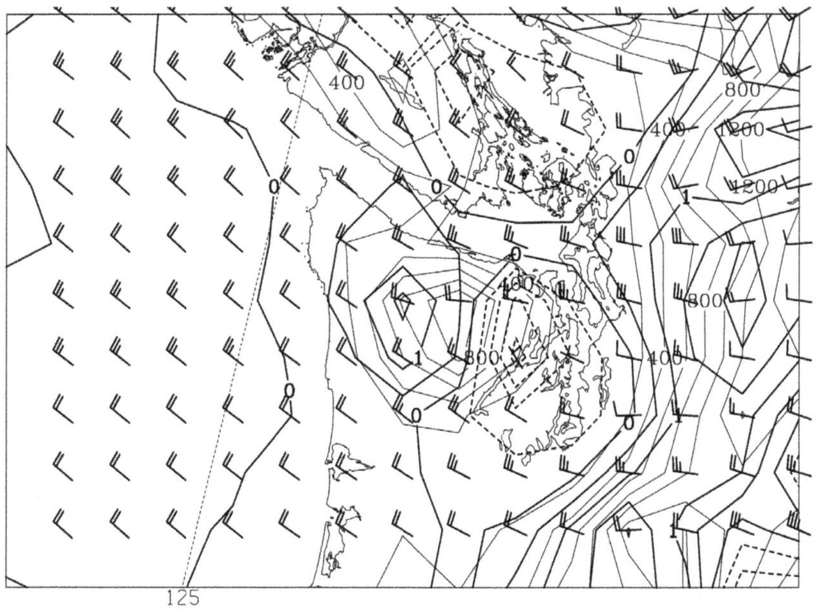

Fig. 8.19 Mesoscale sea-level perturbation pressure (solid/dashed contours every 0.5 hPa) and 850 hPa wind (barbs) near the Olympic Mountains (thin solid lines) for a Puget Sound Convergence Zone case from October 2013.

an anticyclonic eddy on the lee side. For larger islands, these are large-scale features that produce a convergence zone in the lee that extends back toward the island, similar to the Puget Sound Convergence Zone. For smaller islands, this flow around the topography more likely results in Karmen vortices that can extend well downstream from the island.

8.6 Exercises

8.1 Catalina eddies were described in the text as essentially a lee troughing-type response to flow over coastal mountains north of Santa Barbara. Presumably, the same mountain wave considerations examined for initiating a wind reversal would apply here as well. Explain what upstream profile would produce this response for this east–west oriented mountain range. What direction should the flow at 700 hPa be from in order to produce this response? Is that consistent with the typical synoptic pattern for eddy events?

8.2 Calculate the magnitude of the circulation that will result from a 2 hPa low-pressure perturbation centered offshore in the bight region. Assume geostrophy and that the radius of the low center is 200 km. What will the flow speed be if the perturbation is only 1 hPa?

8.3 For Hood Canal wind events, the wind enhancement is the perturbation ageostrophic winds in response to a strong perturbation lee low. If the perturbation pressure response is 4 hPa and it occurs over a 50 km distance, calculate the magnitude of the ageostrophic flow that would occur.

9 Gap Winds and Drainage Flows

Up to this point, we have concentrated on the flow response under blocking conditions and flow over topography where the topography is essentially uninterrupted. However, real topography is characterized by various gaps and valleys, which may result in additional impacts on the flow. In this chapter, the flow response to gap and other breaks in topographic barriers is considered. Gaps and valleys can be wide or narrow and be characterized as flat or sloping. The aim in this chapter is to characterize the flow response to various gap characteristics and background flow patterns. Of interest as well is what happens to the flow once it exits a gap and extends into the offshore waters.

9.1 Description of Gap and Drainage Flows

For purposes in this discussion, gap winds are considered to be those that occur through relatively flat topographic channels, and katabatic or drainage flows occur when the topographic channel has a significant slope. Typically, gap and katabatic flows produce an identifiable low-level wind maxima that extends offshore (downstream) from the opening in the topography. While the offshore response is rather similar, the nature of the forcing and the details of the flow within the gap can vary substantially between gap and katabatic type flows. Figure 9.1 shows the distribution of wind observations offshore and through the Strait of Juan de Fuca along the Washington coast. Notice that the winds are strongest just offshore from the mouth of the Strait. The larger-scale flow is not explicitly depicted but can be inferred from the sea-level pressure analysis. The sea-level pressure contours in Fig. 9.1 suggest that the flow is generally south to southeast across the Strait as well as the entire region. The wind direction in the Strait tends to occur in the direction of the gap orientation or along the gap. Similar high wind regions near the exits of gaps often occur with katabatic flows that extend perpendicularly offshore from sloping valleys and extend out away from the coast.

The depth of these types of flows tends to be rather shallow (below the height of the mountains). This is evident if we examine the flow at 850 hPa (near the top of the topography) for the gap flow along the Washington

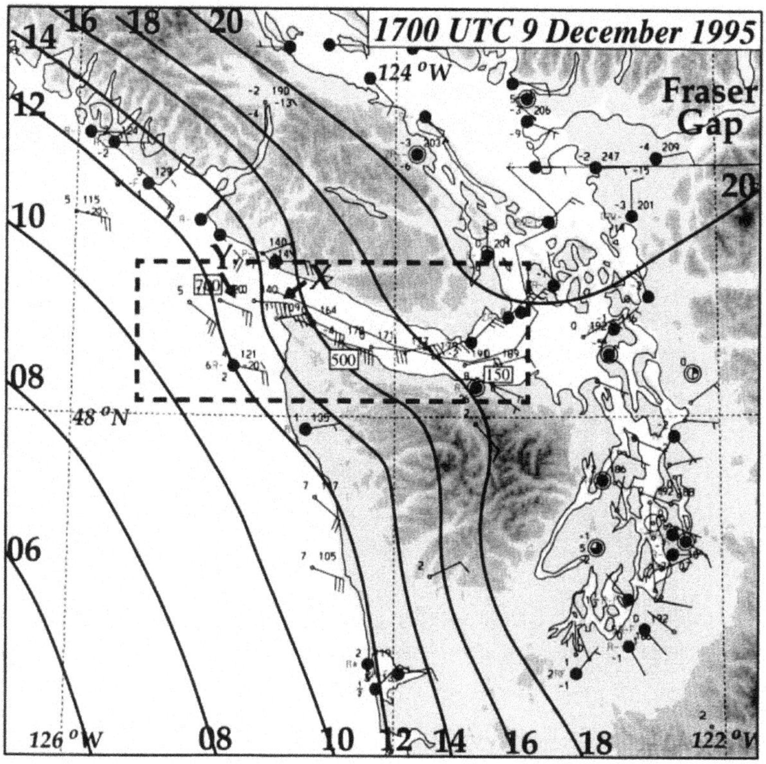

Fig. 9.1 Surface analysis of a gap wind event for the Strait of Juan de Fuca. From Colle and Mass (2000). ©American Meteorological Society. Used with permission.

coast in the cross-section shown in Fig. 9.2. The wind speeds are very low around the 1500–2000 m layer, which is near mountain-top level for the adjacent topography. The 850 hPa flow shows very little evidence of channelling by the gap, with flow directions and speeds similar over the adjacent regions. The near-surface wind speeds within the gap are typically higher than those in the surrounding region. Observations show that wind speeds increase down the length of the gap, reaching their maximum at the mouth or exit from the gap. The low levels tend to be relatively stable with cooler air near the surface, although the static stability need not be any particular value in order to have a gap flow.

Katabatic or downslope flows occur in many mountainous regions and are illustrated by the flow patterns observed in Antarctica shown in Fig. 9.3. While the flow speeds are not shown in the figure, the flow streamlines indicate the downslope nature of these winds. The streamlines concentrate in the sloping valleys to produce distinct outflows across the coast at these valley exits, similar to the gap flow described above.

Katabatic flows are also shallow with the strongest winds just above the surface. The near-surface wind speeds increase through the length of the sloping valley, reaching their maximum near the mouth or exit as well.

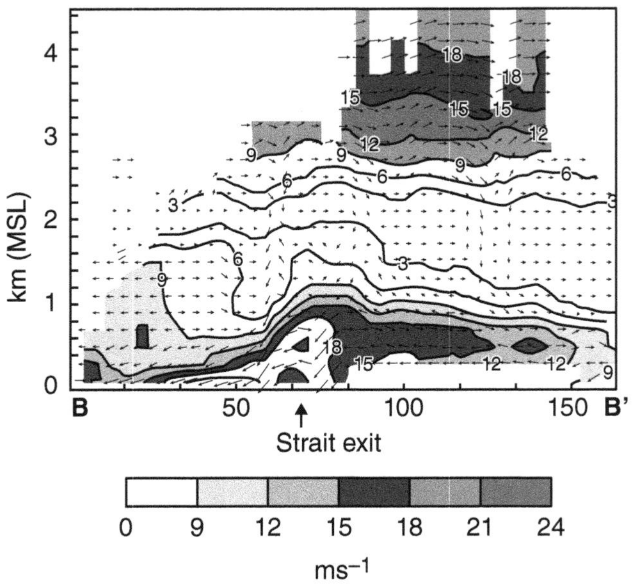

Fig. 9.2 Vertical cross-section of winds through a gap flow based on aircraft observations. From Colle and Mass (2000). ©American Meteorological Society. Used with permission.

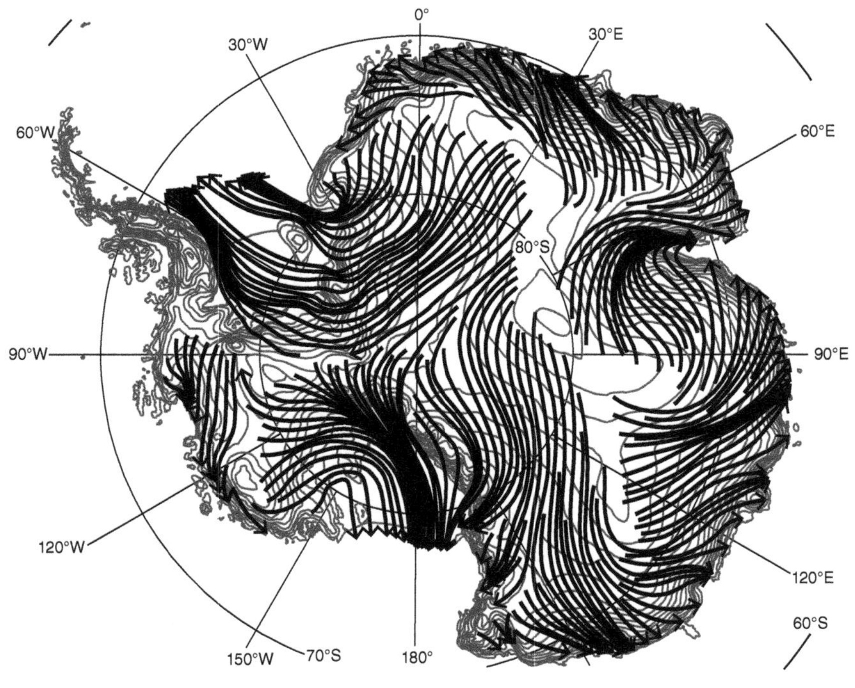

Fig. 9.3 Observed surface flow patterns in Antarctica showing katabatic flows down sloping valleys. From Parish and Bromwich (2007). ©American Meteorological Society. Used with permission.

Although katabatic flows are often cold due to strong surface cooling, the flow as it descends tends to warm adiabatically and can be warmer than the surrounding region. While gap flows can occur under rather weak stratification, katabatic flows have strong static stability due to the surface cooling that forces the flow in the first place. Katabatic flows owe their existence to the sloping topography and can occur along open slopes as well as confined valleys. The strongest flows tend to occur in valleys where the flow is funneled into a single channel, as seen in the Antarctic winds in Fig. 9.3. Weak katabatic flows frequently occur nocturnally in many mountainous regions due to the nighttime surface cooling. Stronger katabatic flows are more typically associated with polar regions, where persistent surface cooling occurs in the winter months due to a lack of solar radiation.

9.2 Synoptic-Scale Pattern for Gap and Katabatic Flows

The synoptic-scale patterns that result in gap and katabatic flows tend to be highly varied. Gap flows are often observed under situations where a cold anticyclone occurs on one side of a gap with lower pressure on the other. This pattern establishes a large-scale pressure gradient across the topographic barrier through which the gap occurs. Figure 9.4 shows an example of one such situation where the cold anticyclone occurs inland over British Columbia with lower pressure offshore. The large-scale pressure gradient is perpendicular to the gap associated with the Strait of Juan de Fuca. The observed winds tend to follow the gap, but their speed is relatively constant. This type of flow is consistent with a balanced flow that aligns with the gap. Trajectories show that the air flow is channeled through the gap but undergoes no particular acceleration associated with the gap. While these types of situations are common for the flow through some gaps, a different dynamic response can be observed when an offshore low-pressure center approaches the coastal topography through which breaks occur, and the pressure gradient rotates to be aligned with the gap. Figure 9.5 shows an example of this type of setup where the low-pressure system off the coast produced gap flow through the Strait of Juan de Fuca. In this case, the winds also blow along the axis of the gap, but distinct acceleration happens between the entrance of the gap to the east and the exit to the west. It is this flow acceleration along the gap that characterizes gap flows. While the basic synoptic pattern varies, the fundamental requirement for the development of a gap flow is simply a synoptic-scale pressure difference across the barrier through which a gap occurs.

Katabatic flows have a more distinct synoptic-scale pattern under which they are likely to occur. As will be examined later, katabatic flows

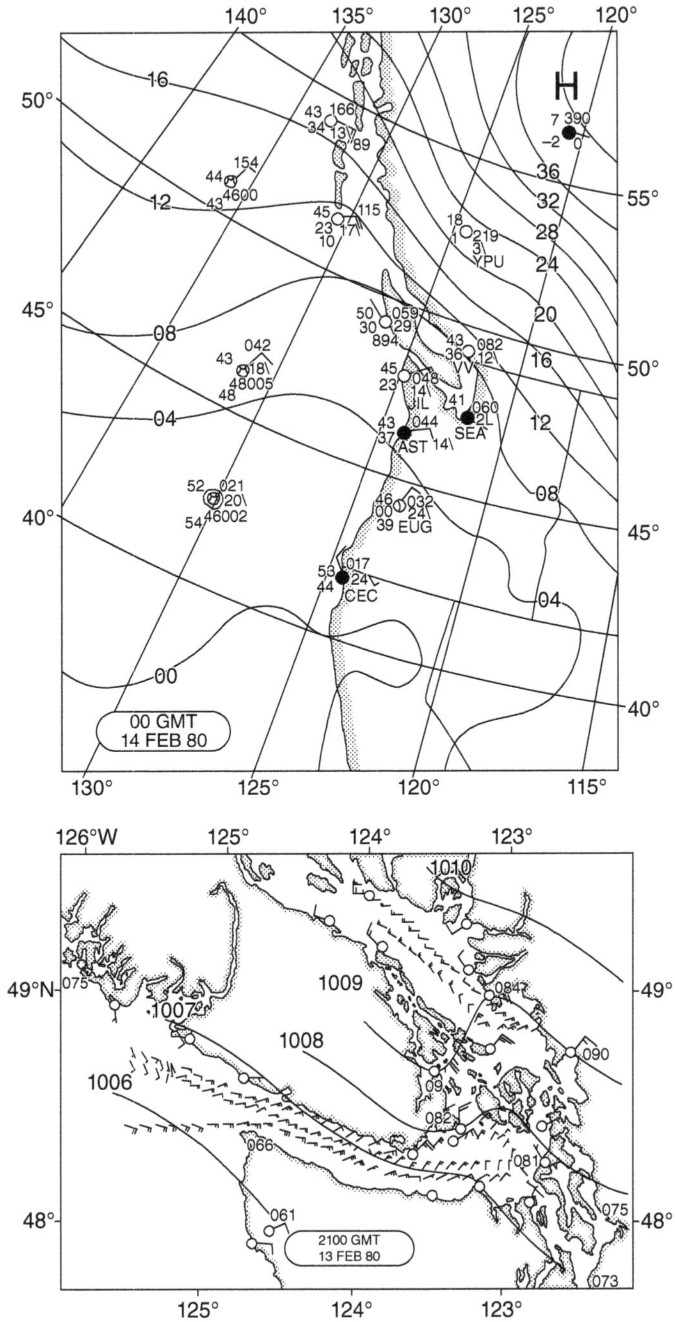

Fig. 9.4 Flow through Strait of Juan de Fuca where pressure gradient is cross-gap. This is characterized as channeled flow. From Overland and Walter Jr (1981). Published in 1981 by the American Meteorological Society.

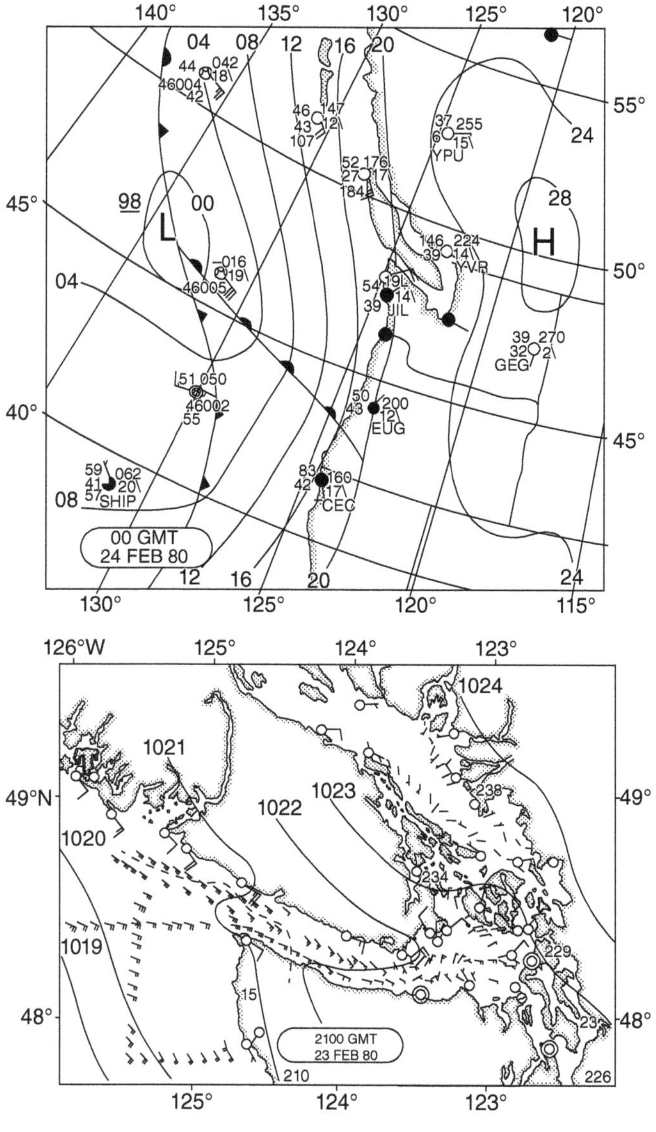

Fig. 9.5

Flow through Strait of Juan de Fuca where pressure gradient is along the gap. This is characterized as gap flow. From Overland and Walter Jr (1981). Published in 1981 by the American Meteorological Society.

owe their existence to strong surface cooling that produces near-surface negative buoyancy along a slope. Cold anticyclones are considerably more favorable for producing the surface cooling and therefore tend to spawn katabatic flows. Weak synoptic-scale winds and surface cooling that are associated with anticyclones setup most katabatic flows. The surface cooling and katabatic flow can occur nocturnally in most any terrain, while persistent strong cooling that occurs over days in polar latitudes can spawn strong katabatic winds. Fundamentally, the synoptic-scale pressure gradient need not point in any particular direction relative to the slope or valley where a katabatic wind occurs, as long as the pressure gradient is sufficiently weak. However, when the pressure gradient force points in the direction of the valley, the katabatic flow need not work against the large-scale pressure gradient.

In addition to the synoptic-scale setup for gap or katabatic flows, mesoscale pressure gradients can contribute to the distribution of winds in a gap and may alter the development of katabatic winds. Several important processes have been identified that contribute to mesoscale pressure gradients in gap flows. First, for gap flows that originate from a cold anticyclone, the low levels on the upwind side of the gap are characterized by a pool of cold air at the surface. The depth of this cold pool contributes to the background pressure gradient through the gap. A pressure difference can be estimated from the perturbation vertical equation of motion (1.5.8c) and assuming hydrostatic balance to yield $\delta p' = \rho g \frac{\theta'}{\theta_o} \delta z$. Assuming a 400 m depth difference in the cold pool, a 12°C thermal difference with the overlying air, and a density of 1.3 kg/m^3, we will get a 2 hPa pressure difference due to this cold pool change in depth. For colder air, the effect can be even greater. This will add to any synoptic-scale pressure difference to enhance the gap flow. In addition, Figure 9.6 shows that as air flows

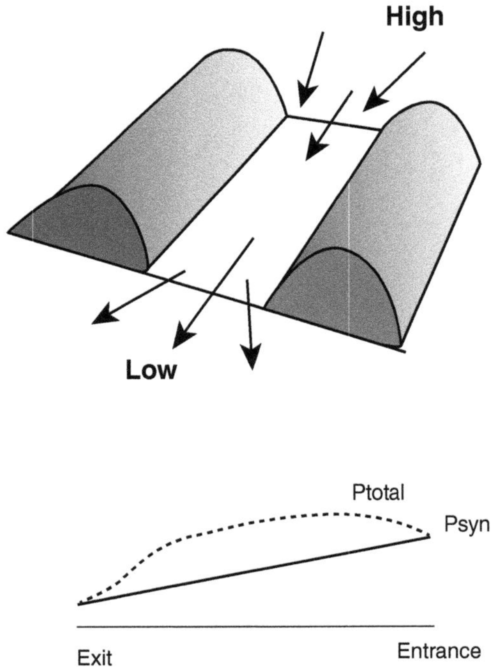

Fig. 9.6 This is the setup that results in gap flow. Top figure shows high pressure on one end of a topographic gap and low pressure on the other end. Arrows show flow converging into the entrance and diverging out the exit, with the strongest winds at the exit. Lower figure shows the pressure distribution through the gap. Solid line shows the synoptic scale background pressure (high at entrance and low at exit). The total pressure is the dotted line, which shows a pressure increase in the gap due to a deepening of the cold air converging into the gap.

toward the gap, the cold pool is dammed up against the topography to develop a mesoscale pressure gradient that opposes the background synoptic scale just upstream of the gap entrance. This damming up of cold air tends to decrease the wind speeds at the entrance of the gap. On the other end of the gap, the cold air is spilling out over a broader area, and this rapid shallowing of the layer tends to develop a mesoscale pressure gradient that enhances the background gradient just downstream from the exit as shown in Figure 9.6. The flow tends to accelerate in this region of the gap producing the strongest wind at the exit of the gap. Another process that has been shown to be important in some gap flows is the development of mountain wave-induced lee troughing at the exit of the gap to enhance the pressure gradient through the gap. The mountain waves and associated lee troughing arise due to the flow over the adjacent topography. The excitation of mountain waves requires that at least the flow profile and stratification across the upper portion of the mountains have a Froude number less than 1. Higher stratification near the mountain top and reverse shear wind profile would favor a larger lee side trough and downslope flow that will impact the mesoscale pressure gradient through the adjacent gap. Mountain waves don't occur in many gap flows but may be very significant for some gap flows.

9.3 Dynamics of Gap Flows

Gap flow dynamics are controlled by the basic horizontal equations of motion. The flow accelerates down the pressure gradient and when unconstrained responds to the Coriolis force to tend toward a quasi-geostrophic balance. At the surface, friction modifies this balance as well to produce some turning of the winds toward low pressure. As the pressure gradient changes, the length scale that describes the distance over which geostrophic adjustment occurs is the Rossby radius $L_r = \frac{U}{f}$. When the pressure gradient is large, implying large U, the length of adjustment is large.

Consider what happens in a channel or gap as the pressure gradient changes from a cross-channel to an along-channel direction. As the pressure gradient rotates, the wind should rotate as well and will when there is no topography. However, in the presence of topography, the wind must rotate to a cross-channel direction to remain in quasi-geostrophic balance. This may or may not be possible depending upon the width of the channel and the stratification of the flow. If the channel is narrow, smaller than a Rossby radius, the distance required for adjustment is longer than the available physical distance across the channel. This will be true for all but the most weakly stratified flows where air easily goes up and over topography. Otherwise, the flow is constrained to flow within the channel or gap and will accelerate through the gap down the pressure gradient. Assuming a uniform pressure gradient through the gap, this response to the pressure gradient will result in the strongest winds at the end or exit of the gap, which is consistent with observations. An important aspect of this dynamic constraint is that it implies that gap width is important to consider and that depending on the background pressure gradient, the width of gaps that will support or produce gap flows will vary and that even rather wide gaps can behave as gap flow when the pressure gradient is rather strong.

The dynamics of gap flow described above can be treated with a rather simple mathematical model of the force balance. The dominant balance is the balance between the pressure gradient through the gap and the acceleration of the wind in the gap. The following equation describes this balance in its simplest form where the transient part of the acceleration is ignored.

$$u\frac{\partial u}{\partial x} = \frac{-1}{\rho}\frac{\partial P}{\partial x} + F. \tag{9.3.1}$$

This equation can be explicitly integrated to get a very simple expression for the wind speed in the gap assuming no friction.

$$u(x) = \left(u(0)^2 - \frac{2\delta P}{\rho}\right)^{\frac{1}{2}}. \tag{9.3.2}$$

These results show that the wind speed will increase through a gap based on the pressure difference between the entrance and exit of the gap. For example, if there is a 5 m/s flow entering a gap and the pressure difference is 2 hPa or 200 Pa, then the wind speed at the exit of the gap will be $18\,\mathrm{m/s} = ((5\,\mathrm{m/s})^2 + (\frac{2*200\,\mathrm{pa}}{1.3\,\mathrm{kg/m^3}}))^{1/2}$. The length of the gap does not enter into this equation as we are integrating the pressure gradient over the entire length. If the gap is short, then this pressure difference implies a strong pressure gradient, and if the gap is long, then the pressure gradient is weak. The end result will be the same overall flow acceleration. As this result suggests, the wind speeds should be rather large at the end of the gap even with a modest pressure difference. The high wind speed is in part because we have neglected any slowing by friction which of course occurs within the flow.

If we include friction, the equation can also be integrated to get an expression for the winds in the gap. The momentum equation can be written as follows by representing the friction as $c_d u^2$:

$$u\frac{\partial u}{\partial x} = \frac{-1}{\rho}\frac{\partial P}{\partial x} + \frac{c_d u^2}{h}. \tag{9.3.3}$$

This can be written as follows:

$$\frac{h}{2c_d}\frac{\partial u^2}{\partial x} = \frac{-h}{c_d \rho}\frac{\partial P}{\partial x} + u^2. \tag{9.3.4}$$

And assuming that $\frac{\partial P}{\partial x}$ is a constant, the equation can be written as follows:

$$\frac{h}{2c_d}\frac{\partial\left(u^2 - \frac{h}{c_d\rho}\frac{\partial P}{\partial x}\right)}{\partial x} = \frac{-h}{c_d\rho}\frac{\partial P}{\partial x} + u^2, \tag{9.3.5}$$

which can be integrated to get,

$$u(x) = \left(\frac{-h}{c_d\rho}\frac{\partial P}{\partial x} + \left(u(0)^2 - \frac{-h}{c_d\rho}\frac{\partial P}{\partial x}\right)e^{\frac{-2c_d x}{h}}\right)^{\frac{1}{2}}. \tag{9.3.6}$$

In this expression, the response is similar to the simple one, except that the wind speed growth is slowed by the friction. Both expressions show that the winds will increase through the gap reaching their largest value at the exit, when this simple dynamic model no longer applies. Using the same conditions as before (5 m/s flow and a 2 hPa or 200 Pa pressure difference), we can determine the gap flow speed including friction. To do this, the drag coefficient, depth of the flow, and length of the gap are also needed. Assuming a drag coefficient of $5x10^{-3}$, a depth of 500 m and a gap length of 100 km (which yields a $\frac{\partial P}{\partial x}$ of 0.002 Pa/m), we can calculate the winds at the end of the gap to be 11.8 m/s.

Both these examples show that the gap exit winds can easily be determined from the pressure difference across the gap. This pressure difference can be estimated initially from the synoptic-scale but actually depends on the mesoscale effects described above. An example of the application of this equation is shown in Fig. 9.7, where both the simple expression derived above and one including friction are compared to actual wind observations. The simple expression leads to a speed increase that follows a quadratic form. This simple expression tends to overestimate the winds, whereas the inclusion of friction helps produce a more realistic wind profile through the gap.

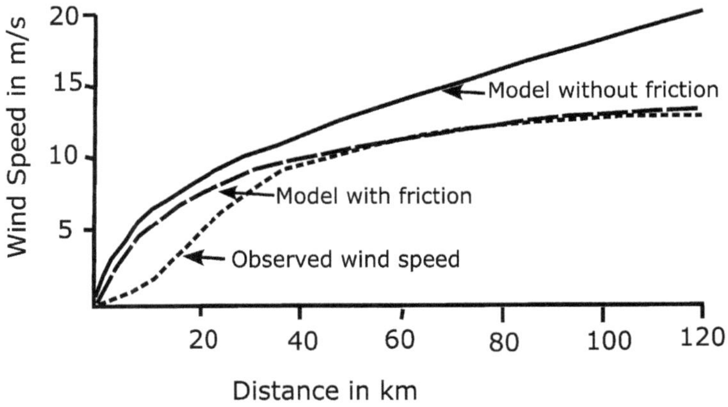

Fig. 9.7 Wind speed variation through a gap based on a simple dynamic model. From Overland and Walter Jr (1981). Published in 1981 by the American Meteorological Society.

9.4 Dynamics of Katabatic Flows

The dynamics that drive katabatic flows are associated with the changes in the near-surface thermal structure that occurs due to surface cooling. Figure 9.8 illustrates the fundamental dynamical forcing of katabatic flows. The stably stratified atmosphere produces the vertical distribution of potential temperature shown by the dotted lines in Fig. 9.8. The isentropes are flat, given no horizontal thermal gradients, and this thermal structure represents the balance between the vertically pointing pressure gradient (labeled P in the figure) and gravity (labeled g in the figure), which consists of hydrostatic balance. When surface cooling occurs, the shallow layer near the ground is modified. Over flat ground, the air cools but the isentropes remain flat which has no impact on the hydrostatically balanced flow. Along the slopes, the cooling results in tilting the isentropes into the vertical as the surface temperature drops most and the cooling goes to zero at the top of the shallow cooling layer marked by the dotted line in Fig. 9.8. This results in tilting the vertical pressure gradient partly into the horizontal as diagrammed in the figure. The small horizontal pressure gradient is not balanced by any other force, and the flow must accelerate horizontally down the slope. This downslope flow is the katabatic wind. The greater the cooling, the more the pressure gradient gets tilted into the horizontal which will drive a stronger downslope flow. This basic forcing explains the primary dynamics but does not give details about the character of the katabatic winds as they flow down the slope.

Insight into the character of katabatic flows can be obtained by considering the more complete dynamic equations. Following Mahrt (1982), the dynamics of katabatic flows can be treated by transforming the dynamic equations to a frame of reference aligned with the slope. This coordinate

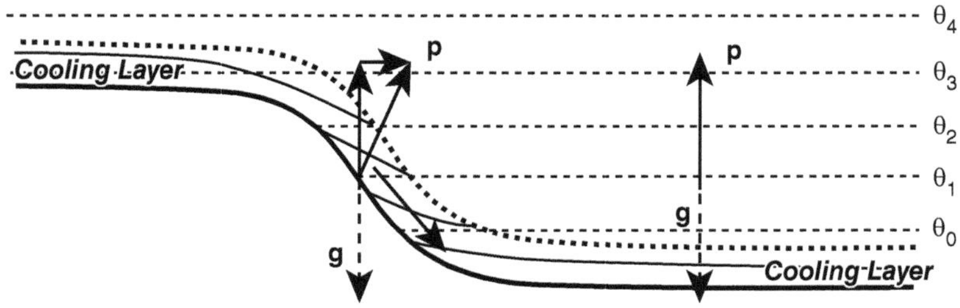

Fig. 9.8 Dynamics that drive katabatic flows. Dashed lines are unperturbed potential temperature contours. Thin solid lines near the surface are the adjusted potential temperature contours after surface cooling has occurred. Heavy dotted line represents the top of the cooling layer. The arrows represent the pressure gradient (P) and gravitational (g) forces over the flat area and over the slope. Arrow along the slope represents the katabatic wind.

shift results in the vertical coordinate, Z, pointing perpendicular to the slope. Given this transformation, the horizontal equation of motion can be written as follows:

$$\frac{du}{dt} = g\frac{\theta}{\theta_0}sin\alpha - cos\alpha\frac{g}{\theta_0}\frac{\partial\bar{\theta}h}{\partial x} + fv - \frac{\partial\overline{u'v'}}{\partial z}, \tag{9.4.1}$$

where α represents the slope angle. The terms in this equation represent that acceleration due to (1) buoyancy, (2) thermal or pressure gradient along the slope, (3) coriolis, and (4) friction. Note that as the angle of the slope goes to zero, the negative buoyancy forcing vanishes (term 1), and the flow acceleration is entirely due to the thermal or pressure gradient, identical to the sea breeze forcing considered previously in Chapter 5. When a slope is present, some of the thermal forcing is mapped into the buoyancy and some into a horizontal thermal or pressure gradient.

Assuming that the katabatic flow can be approximated by a layer of depth, h, through which significant cooling has occurred, the equation of motion can be integrated through the layer to produce an equation governing the evolution of the mean flow in the layer. This flow is considered to be in hydrostatic balance in the vertical and the cold layer is negatively buoyant compared to the adjacent air in the uncooled atmosphere at the same level. This results in a structure where the cool layer along the slope is being forced down the slope by gravity governed by the following equation:

$$\frac{\partial h\bar{u}^2}{\partial x} = hg\frac{\bar{\theta}}{\theta_0}sin\alpha - cos\alpha\frac{g}{\theta_0}\frac{\partial\bar{\theta}h^2}{\partial x} - C_d\bar{u}^2 \tag{9.4.2}$$

Again, the terms in this equation represent that acceleration due to (1) buoyancy, (2) thermal or pressure gradient along the slope, and 3) friction, where the Coriolis term has been dropped. In this equation, the forcing has been integrated vertically through the layer of cooling. The character of the resultant katabatic flow and its associated distribution of winds is a function of which terms tend to dominate in the equation.

To begin to consider the nature of katabatic flows, let's consider the types of flows that result from various force balances. When the acceleration is balanced by the negative buoyancy term, what results is commonly referred to as an advective-gravity flow. This type of flow has the property of the wind speed increasing down the slope due to the extraction of energy from the negative buoyancy. The strongest winds will occur at the bottom of the slope. This is probably the type of flow that typically comes to mind for katabatic flows. A second class of katabatic flows occurs when the force balance is primarily between the friction and the negative buoyancy. This balance implies that any tendency for acceleration by buoyancy is completely offset by friction. Such flows are referred to as equilibrium flows and occur most readily when the layer is shallow, so friction easily acts through the entire depth of the layer. Equilibrium flows arise after a transient flow that results in acceleration up to the equilibrium flow speed.

Another type of flow occurs when the buoyancy, friction and acceleration terms balance. This results in what has been referred to as shooting flow, where the flow on the upper part of the slope accelerates and reaches equilibrium speed part way down the slope. Note that shooting flow is a steady state flow not a transient development. The wind increases partway down the slope, then maintains constant speed for the rest of the way down. The inclusion of all terms, including Coriolis, results in a class of flows known as Ekman-gravity flows. In these flows, the wind turns part way down the slope in response to the Coriolis force to result in flow around the terrain. These types of flows require persistent forcing over several days and are not characteristic of valley flows that are constrained by the topography. Ekman-gravity flows are most frequently observed in Antarctica, where the long-term forcing and broad open slopes can support their development. For the coastal meteorologist, the details of the wind distribution within the katabatic flow are often less important than their speed and subsequent outflow in the coastal region. The key impact here is that even in the absence of a synoptic-scale pressure gradient, a katabatic flow can develop due to cooling and produce a flow out from a valley.

9.5 Offshore Extent of Gap and Katabatic Flows

The outflow of katabatic and gap winds from the coastline is a significant issue for coastal forecasting. How strong will the winds be, and how far offshore will they penetrate? The answer to the first question has been addressed in the preceding sections. For gap winds, the strength depends on the pressure gradient along the gap. For katabatic winds, the strength depends upon the type of flow and how far it descends a slope before flowing out over the coastal waters. For purposes of considering the offshore extent, assume that the synoptic-scale pattern has allowed the development of either a gap or katabatic flow. Gap flow would be associated with a cross-coast pressure gradient (through the gap), and a katabatic flow could potentially have either a cross-coast or along-coast directed pressure gradient.

Let's consider the impact of the orientation of the pressure gradient on the katabatic flow before we consider the nature of the offshore extent. Consider that the katabatic flow is forced due to localized cooling and is not dependent on the synoptic-scale pressure gradient. The winds at the coastline are then the result of this katabatic flow which flows out from the sloping valley perpendicular to the coast. Consider two situations where the synoptic scale pressure gradient is oriented across the coast and along the coast. For the situation of the cross-coast pressure gradient, the katabatic flow is aligned with the pressure gradient and is consequently highly ageostrophic. This results in rather large cross-coast acceleration as the wind must adjust to the pressure gradient. Consequently, a distinct cross-coast flow is observed that is stronger than the flow adjacent to the valley. This can be seen in Fig. 9.9 where the flow near points A and B are strong (around 20 m/s) and directed across the coast. The flow adjacent to these coastal outflows is generally very weak and tends to be in the along-coast direction. For the situation when the pressure gradient lies along the coast, the katabatic flow may be much closer to geostrophic balance with an appropriate direction and a speed that may still be subgeostrophic. In this case, little cross-coast acceleration occurs, and the katabatic wind is hard to distinguish compared to the flow in adjacent regions as seen in Fig. 9.10. The flow near points A and B in this case is again very strong and across the coast. However, the flow adjacent to these points is also across the coast and almost as strong as the outflows from these coastal valleys. This is not to say there isn't a katabatic wind but simply that it is not a distinct feature along the coast. If the katabatic outflow is super-geostrophic, it will decelerate offshore toward the appropriate geostrophic balance. If the katabatic outflow is sub-geostrophic, it will accelerate offshore toward the appropriate geostrophic balance.

Now let's consider the spin down of gap and katabatic flows offshore. The problem of the wind adjusting to the offshore conditions can be considered from several perspectives. First, the slowing of the gap or katabatic wind toward its quasi-geostrophic value could be considered as a frictional spin-down problem. The spin-down depends on the depth of the layer. Using the equation governing katabatic flows of depth h, the spin-down can be treated as a balance between the acceleration term and friction. Using this relationship ($\frac{\partial h\bar{u}^2}{\partial x} = -C_d\bar{u}^2$) and integrating with respect to x, we find that the flow will spin down over a distance given by $L_f = \frac{2*h}{C_D}$, which for a flow that is 500 m deep implies an offshore length scale of 500 km assuming a drag coefficient of $C_D = 2.x10^{-3}$. This probably represents an upper bound on the offshore penetration as frictional spin-down of a deep layer is a relatively slow process. Looking at the problem from the perspective of geostrophic adjustment, the offshore length scale would be given by the Rossby radius. Following Macklin, Lackmann, and Gray (1988), for the topographically forced outflow, the Rossby radius is given by $L_r = \frac{(g'h)^{\frac{1}{2}}}{f}$ where the numerator gives the gravity wave phase speed given their role in causing adjustment. For a 500 m deep flow and assuming a typical g' of 0.33, the offshore distance implied from this perspective is more like 130 km. This distance is more typical of that observed. Figure 9.9 shows that the outflow has decayed by the time it reaches point C (Middleton Island), approximately 200 km from the coast. Both approaches

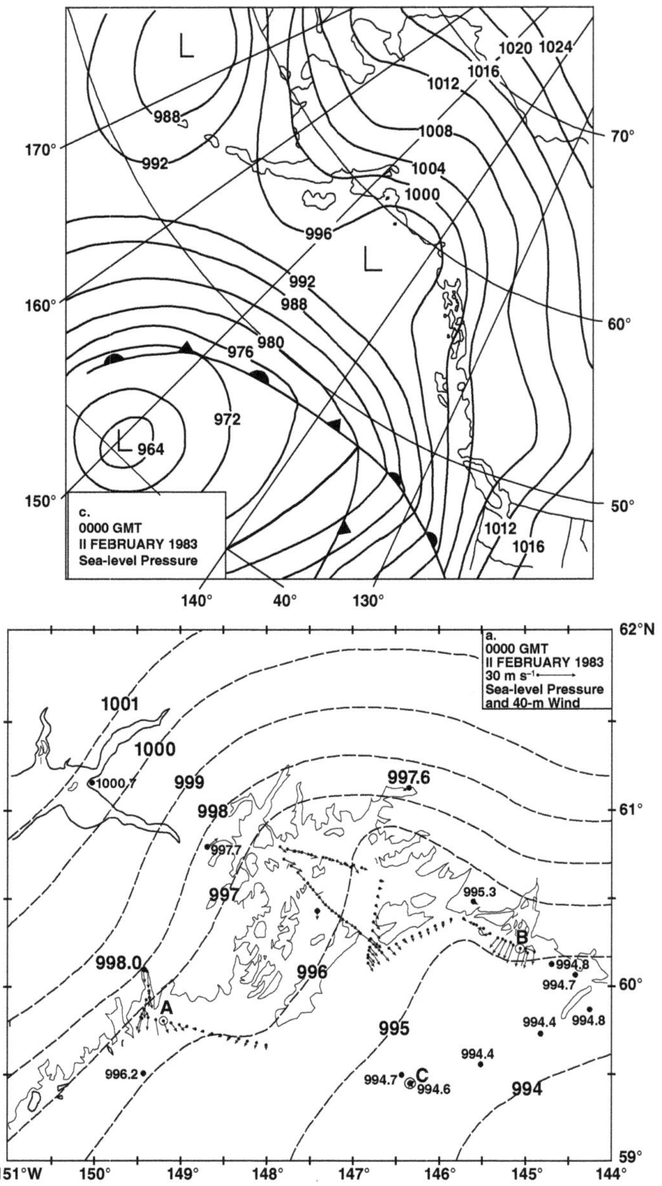

Fig. 9.9 Katabatic wind outflow when the pressure gradient is along the valley. Synoptic scale sea-level pressure analysis is shown in the top figure. Detailed sea-level pressure analysis and aircraft-observed winds are shown in the bottom figure. Isobars are essentially parallel to the coast. From Macklin, Lackmann, and Gray (1988). Published in 1988 by the American Meteorological Society.

to estimating the offshore penetration show a dependence on the depth of the outflow in determining the offshore penetration. The deeper the outflow layer, the further offshore it will go before it can fully adjust. Consequently, shallow flows will not penetrate nearly as far offshore as deeper flows.

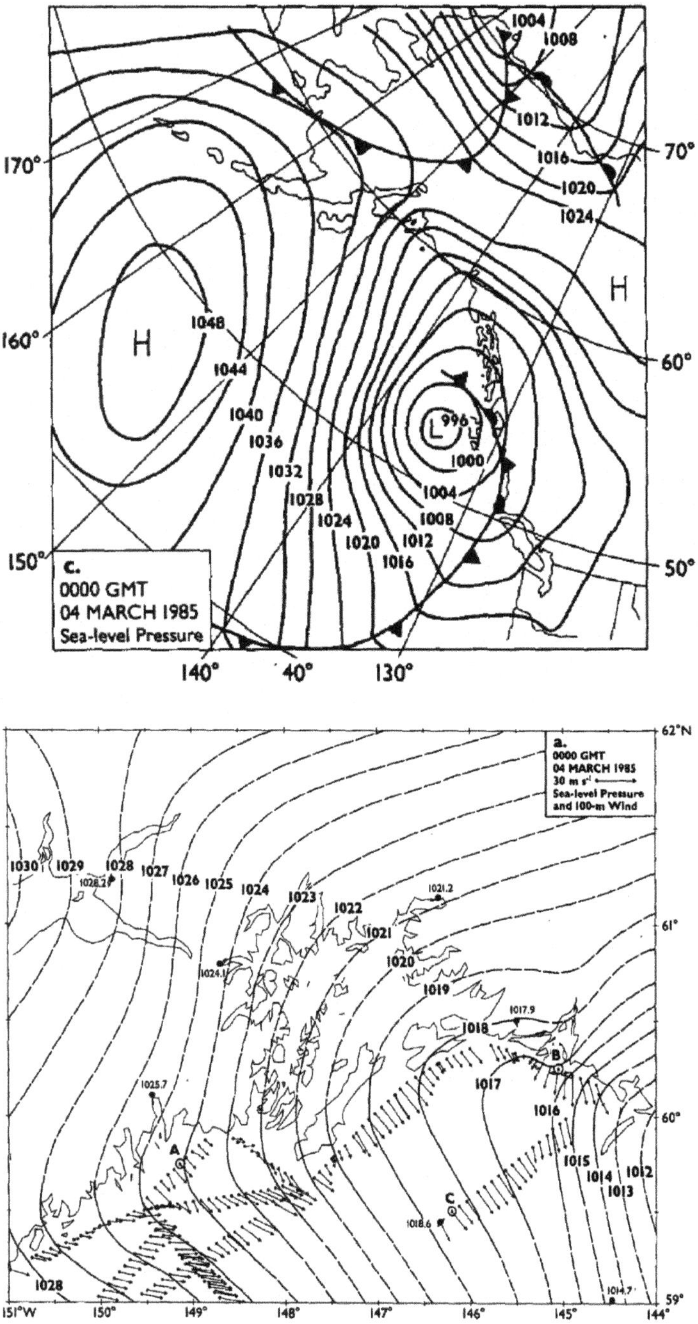

Katabtaic wind outflow when the pressure gradient is cross-valley. Synoptic-scale sea-level pressure analysis is shown in the top figure. Detailed sea-level pressure analysis and aircraft-observed winds are shown in the bottom figure. Isobars are essentially perpendicular to the coast. From Macklin, Lackmann, and Gray (1988). Published in 1988 by the American Meteorological Society.

Fig. 9.10

9.6 Example of Gap Flow

To illustrate the application of these concepts, we will consider the dynamics and evolution of a gap flow through the Chivela pass in southern Mexico. This pass is noted in Fig. 9.11 and produces a relatively regular occurrence of gap flow where air from the Gulf of Mexico flows through the gap out into the Pacific to the Gulf of Tehuantepec. The synoptic evolution is depicted in Fig. 9.11, which shows a cold front in the northern Gulf of Mexico, sweeping down through the Gulf of Mexico that sets up high pressure and cold air on the north side of the gap. Lower pressure occurs to the south on the Pacific side of the gap. The pressure difference is initially near zero as seen in Fig. 9.11 where the pressure is about 1012 hPa on both sides of the gap. The pressure difference increases to about 4 hPa at the peak of the event with higher pressure to the north.

Applying the formula derived for gap flow, the 4 hPa difference would produce a wind speed at the exit of about 30 m/s without friction. If we include friction, the speed would be more like 20 m/s. Figure 9.12 shows a cross-section through the gap of the isentropes and wind speed.. The structure exhibits typical characteristics of gap flows. The pool of cold air is deeper on the entrance side of the gap extends up to above 900 hPa and

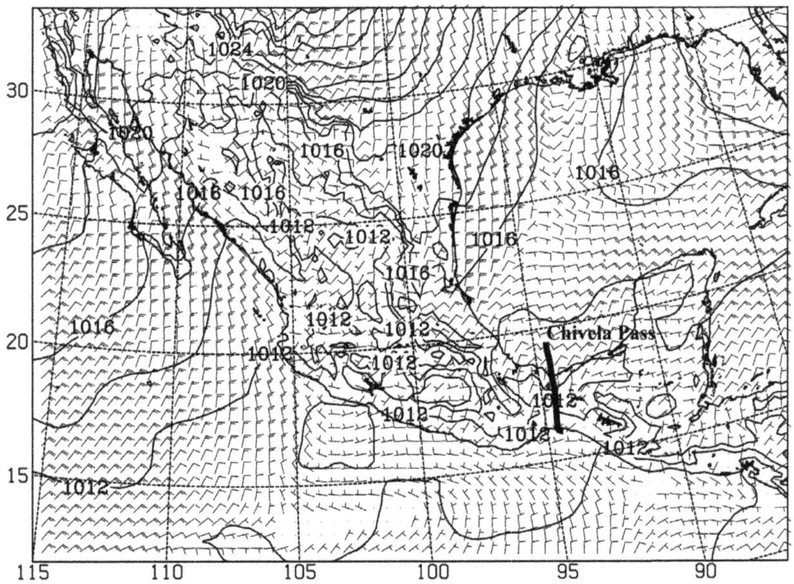

Fig. 9.11 Surface flow pattern for December 06, 2011 at 0000 UTC prior to the start of gap flow over the Southern Mexico and the adjacent oceans (Gulf of Mexico and Pacific). Black contours are sea-level pressure. Black wind barbs are 10 m winds.

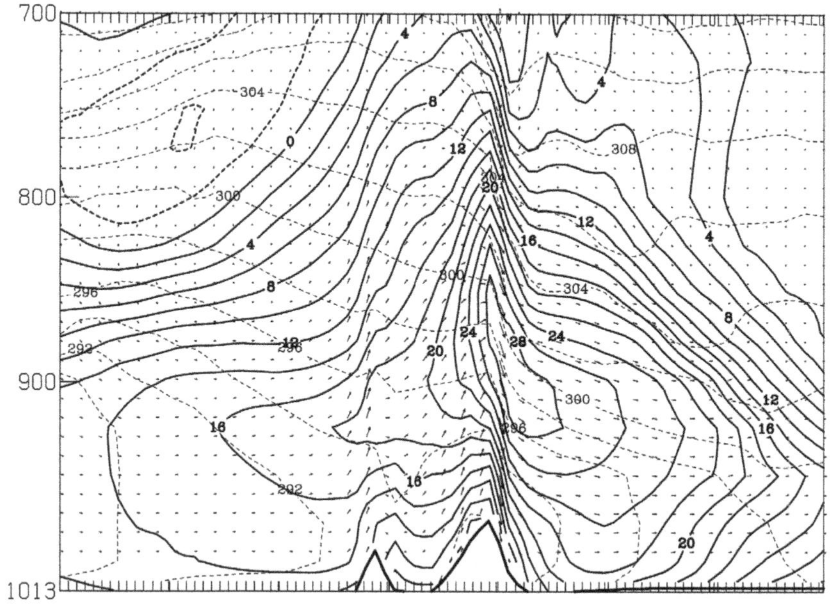

Fig. 9.12 North to south cross-section through Chivela gap at the peak of gap flow on December 08, 2011 0000 UTC. Dashed contours are potential temperature, and solid contours are wind speed isotachs. Arrows are the circuation (vertical and horizontal velocities in the plane of the cross-section).

shallows considerably at the exit to below 950 hPa. Well above the gap near 700 hPa at the top of the figure, the isentropes are nearly flat. The wind speed increases through the gap from 12 m/s upstream to 28 m/s at 925 hPa, with the maximum speed occurring at the exit. The surface winds increase from 12 m/s upstream to about 20 m/s just past the exit. At the mountaintop level near 700 hPa, the flow along the axis of the gap goes to near zero consistent with the flat isentropes and indicating that the gap flow occurs below the topography within the gap.

To examine the offshore response, the near-surface winds and pressure is plotted in Fig. 9.13, near the time of maximum gap flow. Three things become evident about the flow after it exits the gap. First, the flow is highly ageotrophic as it blows perpendicular to the sea level pressure contours. This is expected for gap flows as the pressure gradient driving the flow is along the gap axis. Second, there is a mesoscale high-pressure plume associated with the gap flow. This occurs due to the air in the gap flow being colder and increasing pressure along this flow. In addition, cyclonic and anticyclonic circulations occur adjacent to the gap flow near the exit of the gap. Finally, the flow curves to the right and becomes more quasi-geostrophic after the flow gets further offshore from the gap and the speeds decrease. This occurs west of about 100 W on the figure. The adjustment can be calculated using the two different approaches noted previously. For one, $L_f = \frac{2*h}{C_D} \approx$ 500 km using a depth of 500 m and a drag coeffi-

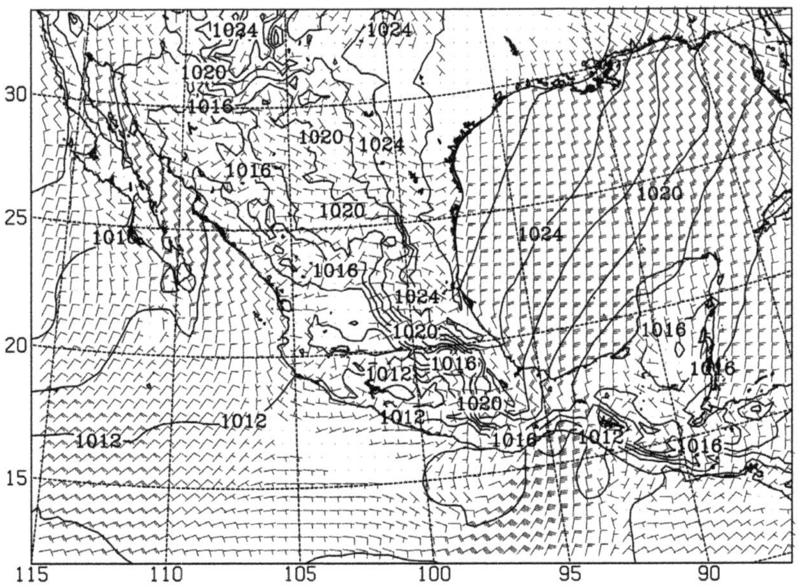

Fig. 9.13 Surface flow pattern for December 8, 2011 at 0000 UTC at the peak of gap flow over the Southern Mexico and the adjacent oceans (Gulf of Mexico and Pacific). Black contours are sea-level pressure. Black wind barbs are 10 m winds.

cient of $2x10^{-3}$. For the other approach, $L_r = \frac{(g'h)^{\frac{1}{2}}}{f} \approx 500$ km using a depth of 500 m and a 6K temperature change across the top of the outflow layer. This geostrophic adjustment occurs over a distance of about 500 km, which is consistent with the Rossby radius for the flow being 500 m deep and the rather low latitude (15N). Both estimates are close to what is observed for the flow to become more geostrophic in Fig. 9.13.

9.7 Exercises

9.1 For a 30 km and 200 km wide mountainous channel, determine the wind speed at which ageostrophic down-gradient channel flow would begin to be observed.

9.2 Consider a 100 km long east–west flat channel between mountain ranges 1500 m high. For a along-channel pressure change of 2 hPa from high (east end) to low (west end) pressure, calculate the following:

- The wind speed at the west end of the channel given an east end wind speed of 2.5 ms^{-1}.
- The geostrophic cross-channel flow assuming the isobars are perpendicular to the channel and $f = 10^{-4}$.

- The magnitude of the lee trough and windward ridge response on the south and north sides of the channel, assuming the stability is relatively low.
- Discuss the impact of the cross-channel flow on the pressure distribution in the channel. What impact does this new pressure distribution have on the winds in the channel?

9.3 Consider the impact of varying the cross-channel geostrophic flow direction in the previous problem. For example, consider the flow going from easterly around to westerly with a low-pressure center propagating north on the west end of the channel. Describe the tendency for the wind in the channel to accelerate down the pressure gradient as opposed to being in geostrophic balance. During what time (i.e., low position) is the greatest acceleration likely to be observed? Does the lee troughing/windward ridging in the channel contribute to the timing and evolution of the flow acceleration in the channel? If so, describe its effect.

9.4 To avoid the strong winds associated with a katabatic outflow along the coast, explain how far offshore you should sail.

9.5 Explain why the strongest winds in a gap tend to occur at the gap exit as the flow accelerates through the length of the gap.

10 Coastal Fronts and Cold Air Damming

Coastal fronts and cold air damming are relatively common occurrences along coastlines in the winter. Coastal fronts form due to the thermal contrast between the cold land and warmer ocean. While studies have focused on the East Coast of North America and the coast of Japan, coastal fronts can occur elsewhere, given the required thermal structure and flow pattern. Cold air damming occurs when cold air is trapped against a mountain barrier and is not strictly a coastal phenomenon. The associated thermal gradient can, in some cases, occur near a coastline to become a coastal front. Although coastal fronts can occur along coasts without topography, cold air damming requires the presence of topography and occurs inland as well as at coastlines. Many cases of coastal fronts are often associated with cold air damming and hence their link in this discussion. As will be examined, coastal fronts and cold air damming occur most readily on the east sides of topography.

10.1 Description of Coastal Fronts

Coastal fronts are low-level baroclinic zones that develop along various coasts due to the thermal contrast established by the land–water interface. Coastal fronts tend to lie parallel to the coast and may move inland somewhat over time. Typically, the synoptic-scale structure does not indicate the pressure of any fronts. This can be seen in Fig. 10.1, which was adapted from a study of New England coastal fronts by Nielsen and Neilley (1990). The sea-level pressure analysis shows the synoptic-scale pattern over New England which generally does not suggest the presence of a front in the region. High pressure dominates over the area and there are no reasons to expect any synoptic scale fronts that are typically associated with low-pressure systems. The synoptic-scale fronts are depicted for the associated lows to the south and west of the area. Figure 10.2 shows the mesoscale surface analysis on the same day, which shows a coastal front over the New England region from one of the cases studied by Nielsen and Neilley (1990). A strong thermal gradient is found near the coast, and a prominent wind shift occurs across the thermal zone similar to other fronts. On the ocean side, southeasterly winds occur in the warm air which are

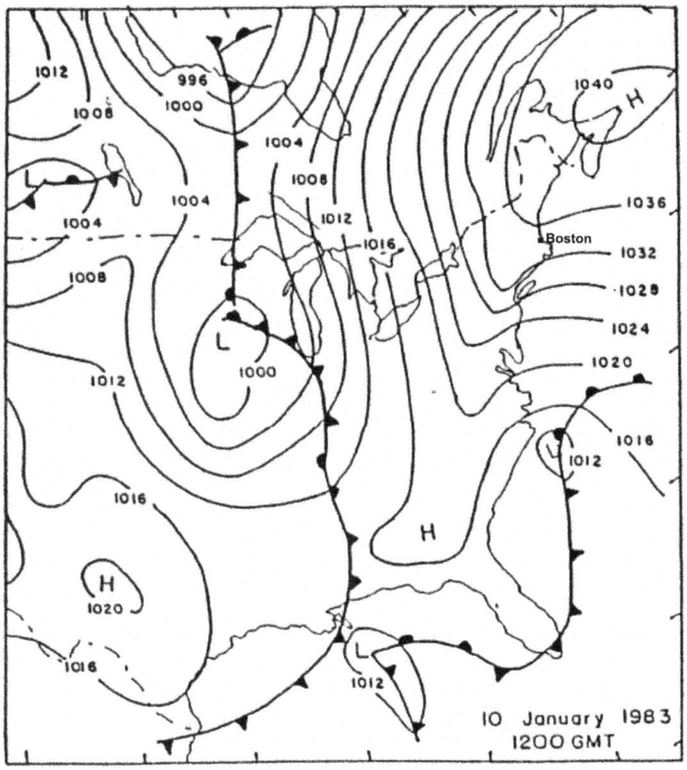

Fig. 10.1 Synoptic pattern under which a coastal front formed along the coast near Boston, marked on the plot. Sea-level pressure contours and surface fronts are shown. From Nielsen and Neilley (1990). ©American Meteorological Society. Used with permission.

perpendicular to the front, while front-parallel northeasterly winds occur in cold air. This suggests that the coastal front is essentially a warm front. This is in contrast to the sea breeze front which acts like a cold front to advect cold air inland.

Nielsen and Neilley (1990) focused on the vertical structure of New England coastal fronts which is generally confined to the boundary layer. The vertical structure of the coastal front from January 10, 1983, is depicted in Fig. 10.3 which shows the shallow nature of these features. The baroclinic zone extends to only about 500 m in the vertical with virtually no thermal gradient above. The vertical structure of the winds shows that the flow is southeasterly in the warm air east of the front as well as the entire region above the front. The low-level winds in the cold air are northeasterly below the strong inversion formed by the baroclinic zone tilting back into the horizontal over the cold air. Streamlines indicate that the warm air is flowing up and over the cold air similar to a warm front. Other cases from the same study had similar characteristics in that they are very shallow features with an inversion extending back over the cold air over the land. This structure is consistent with cold air damming by

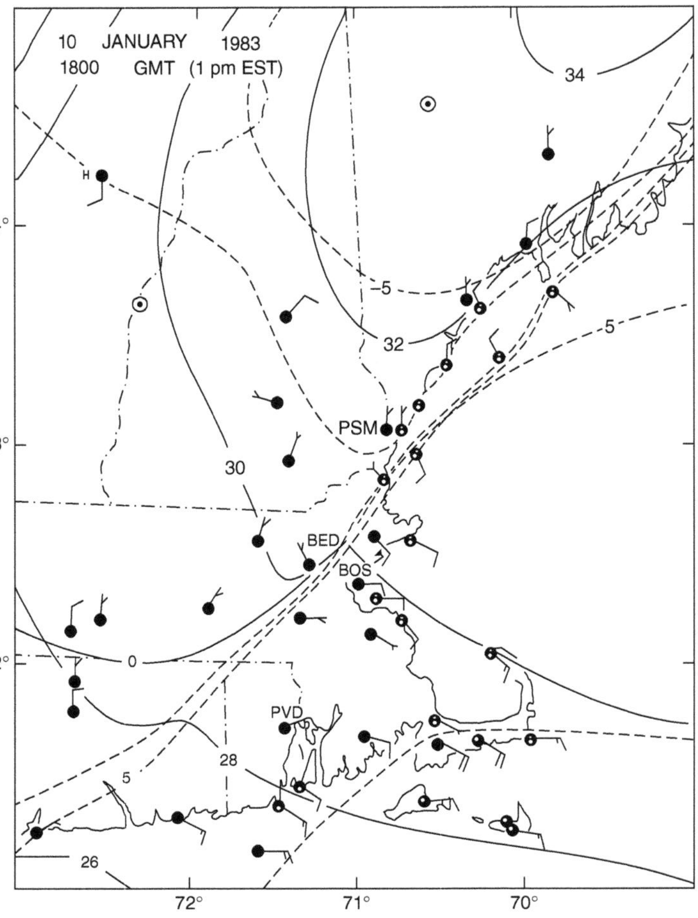

Fig. 10.2 Detailed surface analysis near Boston of a coastal front. Dashed lines are isotherms and solid lines are sea-level pressure contours. Observed wind barbs are plotted such that each long (short) barb represents 5 (2.5) m/s. From Nielsen and Neilley (1990). ©American Meteorological Society. Used with permission.

inland topography, although the formation of the thermal structure and inversion may be tied to the frontal circulation as opposed to blocking by the topography. The shallow nature of coastal fronts has been observed for other coastal fronts as well as these from New England. For example, Raman, Reddy, and Niyogi (1998) found the depth of a coastal front along the North Carolina coast to be shallow with most of the structure occurring below 900 hPa. The shallow vertical structure of coastal fronts indicates their characteristics as a boundary layer phenomena.

Clouds and precipitation are an important aspect of coastal fronts as well. The vertical lift provided by the front itself was a few m/s based on the Nielsen and Neilley (1990) study. This vertical motion is often not sufficient to produce substantial overrunning precipitation due to the shallow

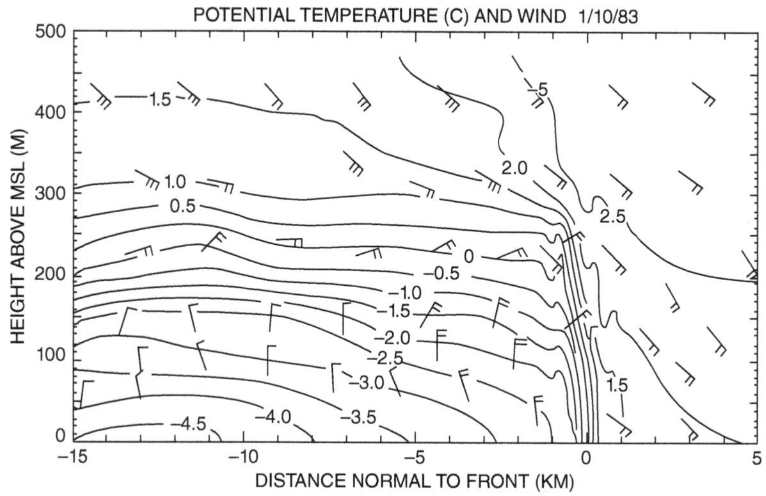

Fig. 10.3

Vertical cross section of winds and potential temperature through a coastal front. From Nielsen and Neilley (1990). ©American Meteorological Society. Used with permission.

nature of the front. However, a local maximum in light precipitation is sometimes observed at the frontal location. More significant is the tendency for the cold air to be characterized by low clouds and fog due to radiational cooling of air moistened by precipitation in the cold air. This tendency to form low-level clouds in the boundary layer behind the front depends very much on the boundary layer dynamics, again highlighting the boundary layer nature of coastal fronts. Another important impact of coastal fronts on clouds and precipitation that occurs because they are wintertime features is their role in producing a distinct rain–snow boundary for precipitation that is often originated above the coastal front in an approaching larger-scale low-pressure system. The dynamics/thermodynamics that lead to differing precipitation types will be examined in Section 10.4.

Coastal fronts are thermally driven circulations governed by different processes than previously considered for sea–land breeze circulations. If we compare the characteristics of the coastal front and the sea breeze front, the coastal front acts like a warm front with the front perpendicular flow in the warm air, while the sea breeze front has front perpendicular flow in cool air over the water to act like a cold front. Land breeze fronts would have the thermal pattern similar to the coastal front but also tend to behave like a cold front with front perpendicular flow occurring in the cold air in an offshore direction. Thus the dynamics driving the coastal front depend on other factors beyond just the thermal gradient, forcing the circulation as observed in the sea/land breeze.

10.2 Cold Air Damming

While coastal fronts can form in the absence of cold air damming, many coastal fronts are tied to the process of cold air damming. Cold air damming occurs when cold air at low levels is trapped against topography due to low-level flow blocking. The typical scenario by which cold air damming is established occurs when cold air is advected south to the east of a mountain range. This cold advection occurs on the east side of an anticyclone after a cold front passes east- and southward as seen in Fig. 10.4 for the Eastern part of North America. The cold air is often rather shallow and can be further enhanced through radiational cooling at night.

The damming begins as the high-pressure center shifts to the east to turn the flow toward the topography which can undergo blocking by the inland mountains. The flow near the surface initially produces the surge of cold air east of the mountains with warm air moving northeast to the west of the mountains as depicted in Fig. 10.4. The cold air east of the mountains is usually characterized by strong static stability because the coldest air occurs at the surface. As the high shifts east, the flow east of the mountains

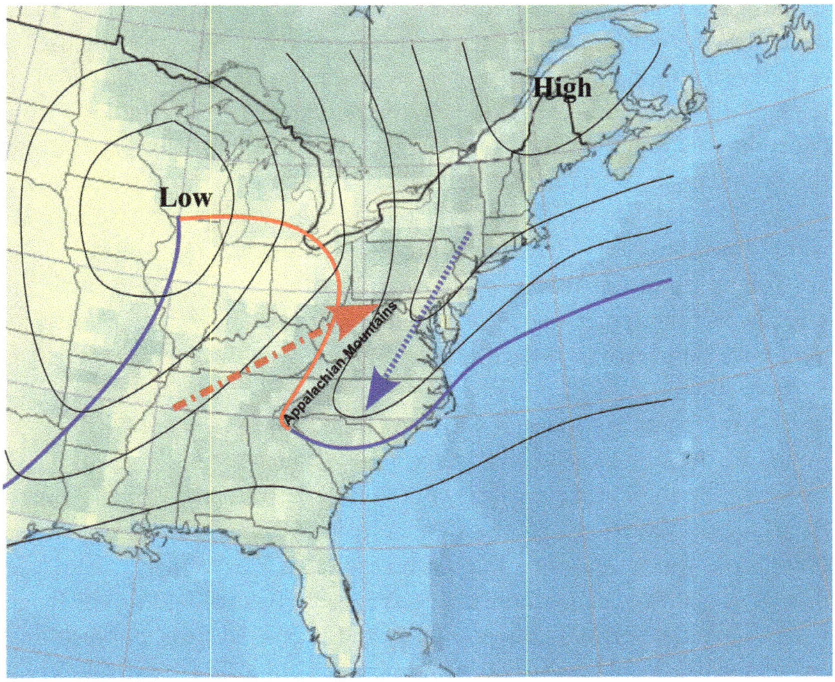

Fig. 10.4 Example of the synoptic pattern associated with cold air damming east of the Appalachian Mountains. Cold air flows out of a surface high to the north as noted by the blue dashed arrow, and warm air advects northward to the west of the mountains as noted by the red dashed arrow. The warm air is advected over the cold air setting up a strong inversion east of the mountains.

and off the coast turns to come from a more easterly direction toward the mountains. The warm surface flow west of the mountains, associated with the surface cyclone to the west, starts to shift to a more southerly direction as the low moves east. As the flow becomes more southerly, it can start to cross the mountains to further increase stability east of the mountains. This pattern of cold air flow toward the mountains from the east and warm air flow over the mountains from the west creates strong low-level static stability in the cold air mass that results in flow blocking of easterly flow. The blocked flow can not flow over the mountains, and so within a Rossby radius of the mountains, the flow is turned toward the low pressure to the south. The winds in the blocked region often never really shift from their northerly to northeasterly direction prior to the high moving eastward. The flow further to the east turns easterly, and as it encounters, the blocked flow is forced to rise up and over the blocked flow. This establishes the cold air damming which can be further strengthened by the easterly flow advecting warm air over the cold air and increasing the static stability above the cold air. The cold air damming is often quite persistent and has a tendency to re-enforce itself. The along mountain flow from the north or northeast will continue to advect cold air into the region to maintain or strengthen the static stability and maintain the flow blocking. The cold air damming decays only when relatively strong synoptic forcing begins to erode the stable layer and the surface cold advection weakens.

This pattern of development of cold air damming due to thermal advection is the primary mechanism to create the cold air east of the mountains. This represents classic cold air damming where things are largely driven by the synoptic-scale flow. Low-level cold air can also be produced or enhanced by evaporation of precipitation in the low levels east of the mountains. This diabatically driven cooling produces what is known as hybrid cold air damming. This process depends strongly on the dryness in the low levels and the favorable synoptic pattern to produce precipitation east of the mountains.

While flow blocking can and does occur on either side of a mountain range, the occurrence of cold air damming is essentially limited to the east side of the mountains. Consider the situation that can occur to the west of a mountain range. Cold air can be advected over the region equally well with a passing cyclone to the north. This establishes low-level static stability due to the vertical structure of the cold air, and it maximizes near the surface. Now to produce flow blocking on the west side of the mountains, the flow must develop a westerly component, which occurs with an approaching low-pressure system. This type of blocking does occur and will be examined more thoroughly Chapter 11. In the blocked region, an along-barrier flow toward low pressure to the north will develop. Given generally warmer temperatures to the south, the barrier flow will advect warmer air northward which tends to weaken the static stability. The static stability reduction will eventually be sufficient for the flow to become unblocked, and any semblance to cold air damming is gone.

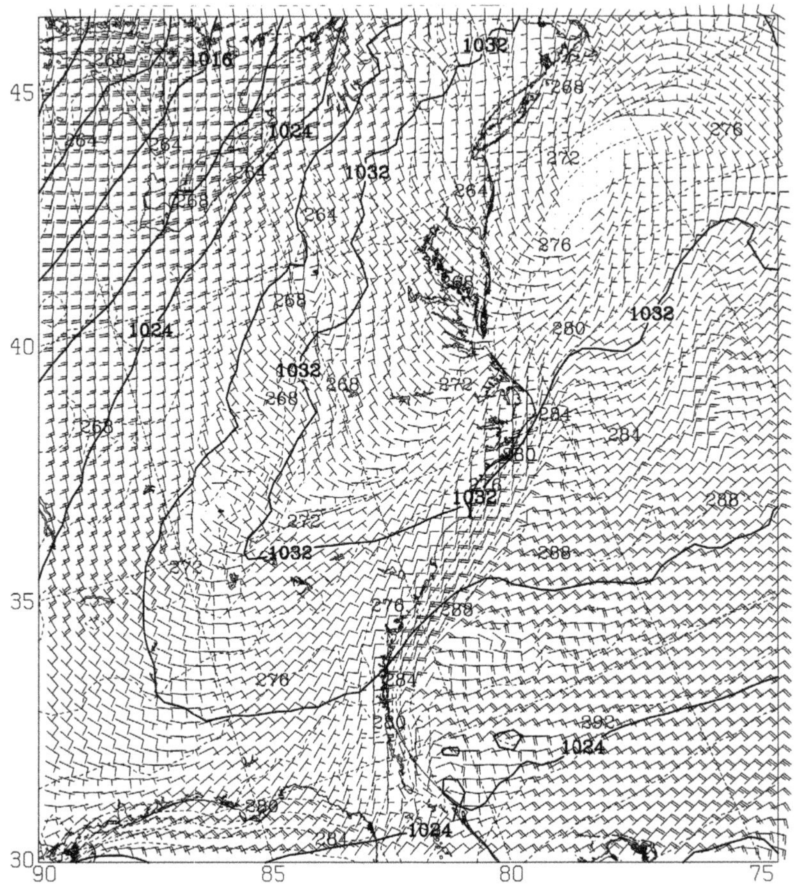

Fig. 10.5 Occurrence of coastal front east of a region of cold air damming. The sea-level pressure (solid thick contours, every 4 hPa), the 1000 hPa isentropes (dashed contours, every 2°K, and 10 m winds are shown. The thermal gradient is set up between the cold air over land, and warm air being advected toward the blocked region over the ocean.

The cold air damming sets up a boundary between the cold air and the warmer air flow from the east. This thermal boundary is also characterized by a wind shift due to the flow blocking and can become essentially a coastal front when it occurs in the proximity of the coast. This pattern is depicted in Fig. 10.5 where a coastal front occurs on the east side of the cold air damming to the east of the Appalachian mountains. The sea-level pressure pattern in Fig. 10.5 shows geostrophic flow from the northeast to east off the coasts of North and South Carolina, and a distinct region of high pressure over the eastern parts of those states. This high-pressure region occurs because of the low-level cold air dammed up in this region. The location of the coastal front is dictated by the Rossby radius associated with the flow blocking. For this case, the Rossby radius can be calculated using $L_r = \frac{Nh}{f}$, with $N = 0.02\,\mathrm{s}^{-1}$, a Coriolis parameter of

$8.34x10^{-5}$ and a height of the mountains to be $h = 1{,}500$ m, to get a blocking distance of about 375 km which is comparable to the distance from the mountains to the coast in this area. Similar features occur upstream of blocked flow in interior regions as well. For example, Dunn (1987) has documented the occurrence of cold air damming east of the Rocky Mountains in Colorado. The cold air damming produces an associated frontal zone and convergence at the leading edge of the damming region. In interior regions like this, we would not call them coastal fronts, but the dynamics that drive the blocking and cold air damming are the same.

10.3 Synoptic Patterns and Mechanisms for Formation

The development of fronts is associated with flow deformation and frontogenesis, which typically occur in the warm and cold frontal regions of cyclones. Frontogenesis is the dynamical process by which flow deformation acts to increase the horizontal temperature gradient. In its simplest form, the horizontal frontogenesis is given by the following equation:

$$\frac{d}{dt}\left(-\frac{\partial \theta}{\partial y}\right) = \frac{\partial u}{\partial y}\frac{\partial \theta}{\partial x} + \frac{\partial v}{\partial y}\frac{\partial \theta}{\partial y} + \frac{\partial \omega}{\partial y}\frac{\partial \theta}{\partial p} - \frac{\partial \dot{Q}}{\partial y}. \tag{10.3.1}$$

In this form of the frontogenesis equation, the coordinate axes have been rotated so that y points across the front toward the cold air and x is in the along front direction. The terms on the right-hand side represent the processes that act to change the thermal gradient. Shearing deformation, confluent or stretching deformation, tilting by differential vertical motion, and diabatic heating gradients can all act to strengthen or weaken a horizontal thermal gradient. Shearing deformation acts to rotate a thermal gradient so that the isotherms lie along the shear axis, thereby increasing the thermal gradient across the shear axis. Confluent or stretching frontogenesis occurs when the flow comes together across the thermal gradient to compress the isotherms across the confluent axis. Tilting acts to rotate a vertical thermal gradient into the horizontal increasing the gradient in the horizontal. Diabatic heating/cooling acts to create or destroy the thermal gradient through differential heating/cooling in the horizontal. In cyclones, the deformation is primarily the result of the geostrophic flow patterns but can be aided by ageostrophic forcing as well. If we consider the typical synoptic pattern under which coastal fronts are found, the low-level geostrophic flow is not usually frontogenetic, and so we must consider the contributions of the ageostrophic parts of the flow. Figure 10.5 shows the near surface (10 m) winds and the near-surface (1000 hPa) isentropes. A relatively strong thermal gradient can be seen along the North and South

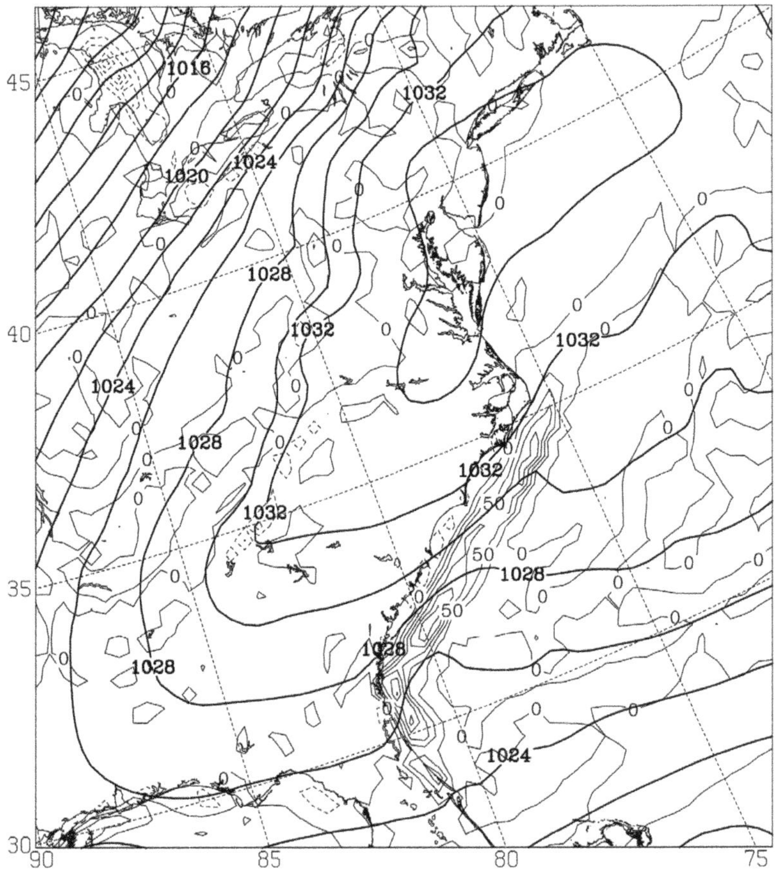

Fig. 10.6 Example of coastal frontogenesis where ageostrophic flow causes frontogenesis. Sea-level pressure isobars are the solid thick contours, every 2 hPa and 1000 hPa frontogenesis is shown as thin solid/dashed contours, every 25°C/100 km/day.

Carolina coasts with the surface winds showing a highly confluent flow in the region of the thermal gradient. This character of the low-level flow in the region of a thermal gradient implies that frontogenesis is occurring. Figure 10.6 shows the calculated horizontal frontogenesis, confirming the strong frontogenesis along the coast. The low-level flow that produced this frontogenesis is highly ageostrophic, particularly over land where the wind is almost perpendicular to the isobars. The geostrophic flow as inferred from the sea-level pressure isobars is largely frontolytic (not shown) in the region where the coastal front forms. The geostrophic flow generally shows confluence seaward of the inverted pressure trough located on the warm side of the thermal gradient. While this confluent flow could be frontogenetic, it occurs away from the thermal gradient and so no frontogenesis is found. On the inland side of the inverted pressure trough, the geostrophic flow is difluent, which does not produce frontogenesis either, even though it is in a region with a strong thermal gradient. When the ageostrophic part of the flow is added back in as shown in Fig. 10.5, largely due to friction over the land part of the flow, the same region becomes confluent and thus frontogenetic as shown in Fig. 10.6. Several mechanisms have been identified by which the necessary low-level frontogenesis is established to create a coastal front. These mechanisms occur under different synoptic regimes that are described below.

The first type of coastal frontogenesis is associated with the movement of a cold anticyclone across the coast to produce on-shore geostrophic flow. In these situations, a cold anticyclone typically occurs after a cold front passes and the geostrophic flow might be offshore directed or even essentially coast parallel. As the anticyclone moves offshore, the anticyclonic circulation will produce a region of onshore directed geostrophic flow. As previously noted in Chapter 2, onshore-directed geostrophic flow will produce coastal convergence, which can become frontogenetic given a cross-coast thermal gradient. This synoptic pattern is the most frequent type to spawn coastal fronts and to produce cold air damming.

The synoptic pattern for the New England area is depicted in Fig. 10.7 for this type of coastal front development. In this case, a cold anticyclone is centered over northern New England 3 hours prior to coastal front formation (Nielsen (1989)). The anticyclone shifts east over Nova Scotia by 3 hours after the coastal front has formed. This slow movement to the east results in the flow around the high to go from northerly to southeasterly over the coastline of Massachusetts to southern Maine. This cold anticyclone typically occurs after the passage of a cold front and the associated cold air outbreak that follows. During the cold air outbreak, cold air is advected across the region and out over the ocean. With ocean surface temperatures warmer than the air, surface heat fluxes warm the air over the water to result in a broad thermal gradient across the coast. To generate the coastal front, this thermal gradient needs to be intensified along the coast.

To consider how this intensification happens, let's examine the detailed evolution of a particular case. The upper left panel in Fig. 10.8 shows

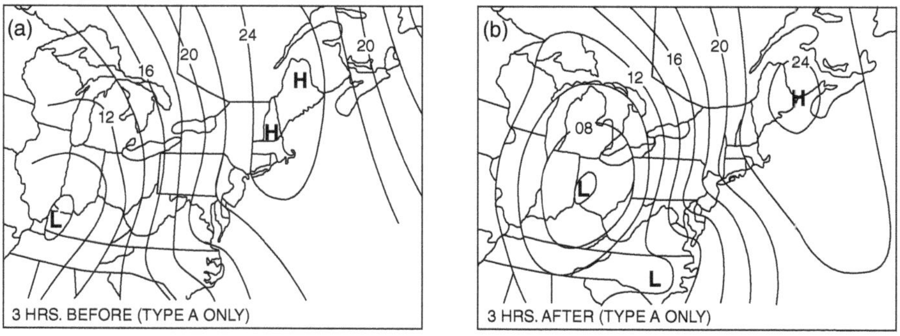

Fig. 10.7

Composite synoptic pattern associated with Type A coastal frontogenesis. Time 3 hours prior to coastal front formation shown in (a) and 3 hours after shown in (b). From Nielsen (1989). ©American Meteorological Society. Used with permission.

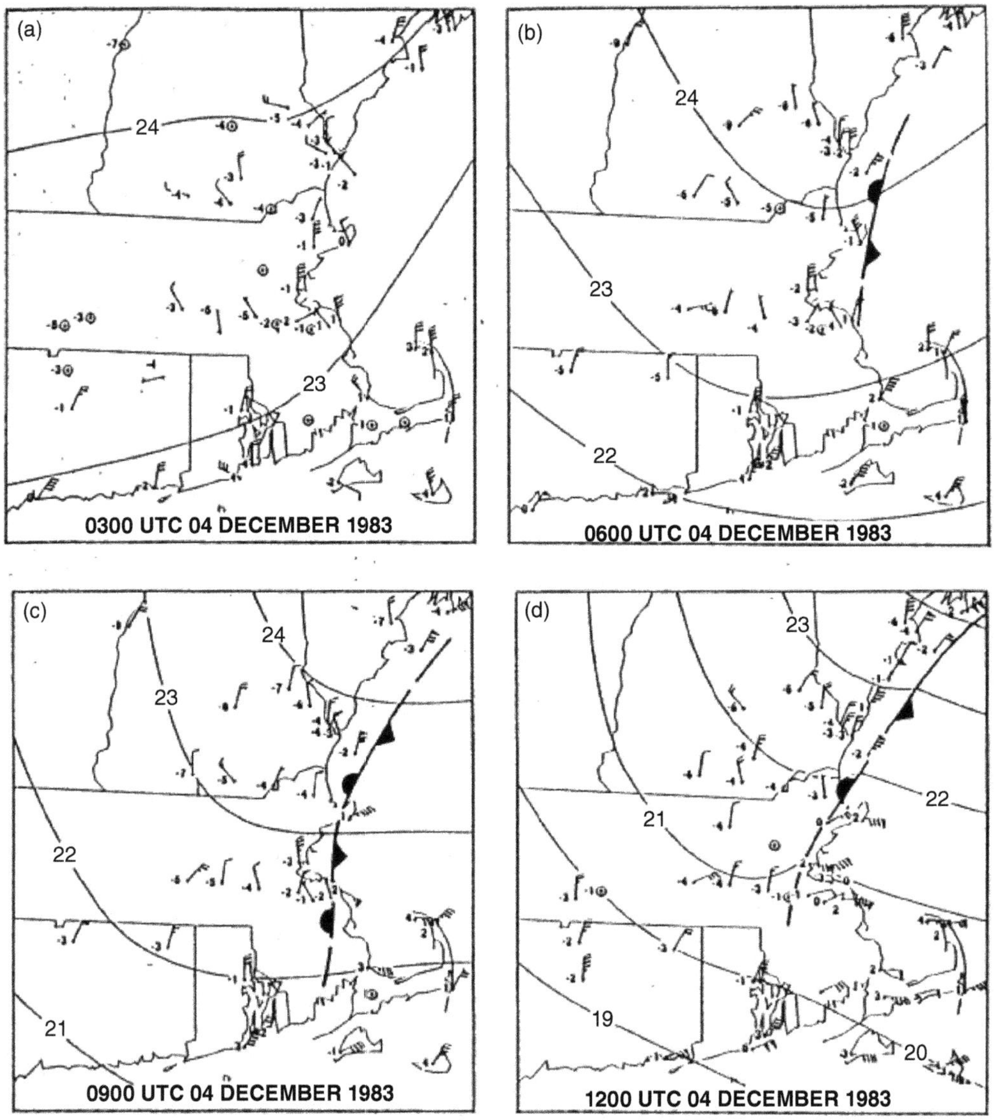

Fig. 10.8 Example synoptic pattern associated with Type A coastal frontogenesis event. Synoptic evolution from (a) high pressure inland and offshore flow, (b) initial front formation offshore, (c) strengthening front and extension along coast, and (d) fully developed front with some inland penetration. From Nielsen (1989). ©American Meteorological Society. Used with permission.

northerly to northwesterly winds over New England and isobars across the coast with lower pressure offshore. This is the final stage of cold advection across the coast. As shown, the geostrophic wind is essentially parallel to the coasts of Maine and New Hampshire. Three hours later, in the upper right panel of Fig. 10.8, the high to the north has shifted eastward so that the geostrophic flow is more northeasterly with a slight onshore component. As the geostrophic wind shifts in this manner, differential friction over the land-ocean boundary tends to force

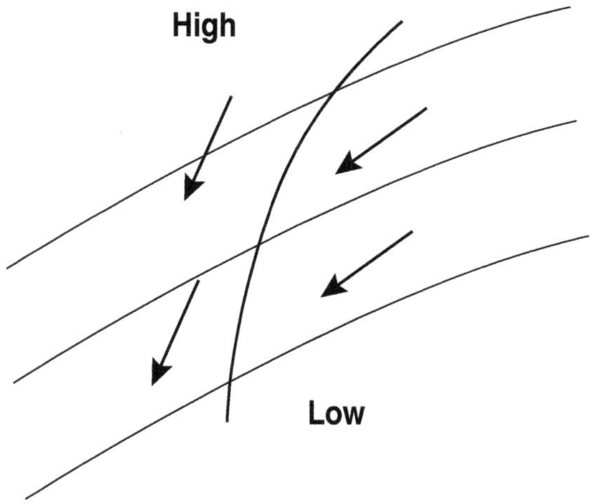

Fig. 10.9 Effect of friction causing coastal convergence. Winds over the land (left side) turn more toward low pressure than those over the water (right side).

confluence at the coast. This is illustrated in Fig. 10.9, which shows how the friction differences of land versus water lead to confluence at the coastline. The surface winds over the land are turned more directly down the pressure gradient than those over the water due to the differing surface roughnesses of the land versus the water. Note that as we saw in Chapter 2, the degree of wind turning toward low pressure depends upon the surface roughness (land use) as well as the depth of the boundary layer. A shallow boundary layer will result in greater frictional turning, and hence the cold stable boundary layer over the land will have greater turning than if it were a deeper layer. This differential frictional turning results in coastal confluence, which is evident in the observed surface winds in the upper right panel of Fig. 10.8. Given a cross-coast thermal gradient, this confluence becomes frontogenetical. As this confluence develops, the coastal front tends to form first along the New Hampshire coast and then extend more north and south as the high continues to move offshore as shown in the bottom panels of Fig. 10.8. This type of coastal frontogenesis has been referred to as type A by Nielsen (1989) and depends upon the differential heating by surface fluxes to establish the thermal gradient and differential friction to force confluent frontogenesis. The lack of turning of the winds over the land areas as the high moves east (offshore) is an aspect of flow blocking and cold air damming processes that help to strengthen the coastal front. Similar processes are observed further south along the Carolina coastline as shown in Fig. 10.6, here the high is positioned further south but produces geostrophic onshore flow over the Carolina coast. Coastal fronts in over regions around the world show similar characteristics with different synoptic setups due to the specific coastal orientations. Similar patterns of coastal frontogenesis have been observed in Japan over the Kanto plain near Tokyo. In this region, the primarily east–west orientation of the coast requires that the anticyclone shift to the southeast of Japan to give southeasterly geostrophic flow over the Japan coastline. The key component in all of these cases of coastal frontogenesis is to establish frictional confluence along the coast to produce frontogenesis.

Nielsen (1989) identified another synoptic pattern for the New England area that can also spawn a coastal front due to flow blocking. This type of coastal front occurs when low-pressure system tracks north and east to the west of New England as shown in Fig. 10.10. For this synoptic scale pattern, the low-pressure region dominates, and the New England area is situated to the east of the low in a region of broad southerly to southeasterly flow. In this situation, the air at low levels is usually not that cold, and the region is experiencing warm advection to the east of the low. The warm advection tends to be strongest just above the surface, where the winds are less impacted by friction and the thermal gradient remains strong. This differential warm advection tends to increase stratification over time as the system moves north and east. This differential warm advection is depicted in the hypothetical profile shown in Fig. 10.11. The warm advection at the top of the boundary layer increases the temperature by 4° per day, while the surface increase is only 2° per day. This vertical temperature difference leads to an increase in the Brunt-Vaisala frequency of 0.036 s^{-1} per day for a 500 m deep layer. This can produce about a 15%–25% increase in the Brunt-Vaisala frequency in 6 hours for typical stratifications. If

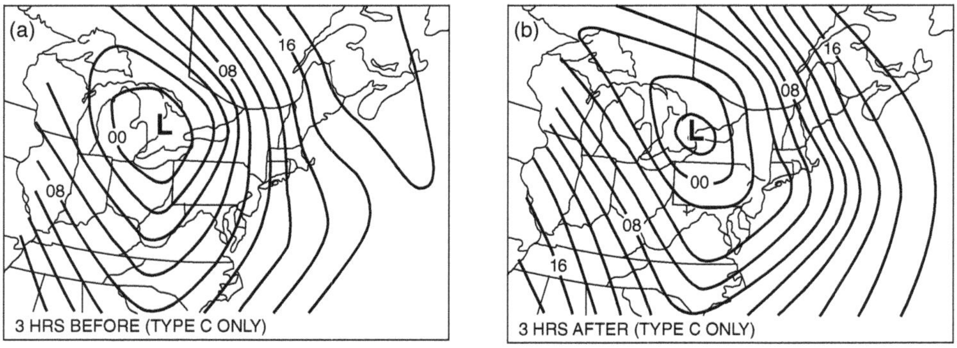

Fig. 10.10 Composite synoptic pattern that leads to type C coastal frontogenesis. Sea level pressure pattern (a) 3 hours prior to coastal frontogenesis to (b) 3 hours after the front forms. From Nielsen (1989). ©American Meteorological Society. Used with permission.

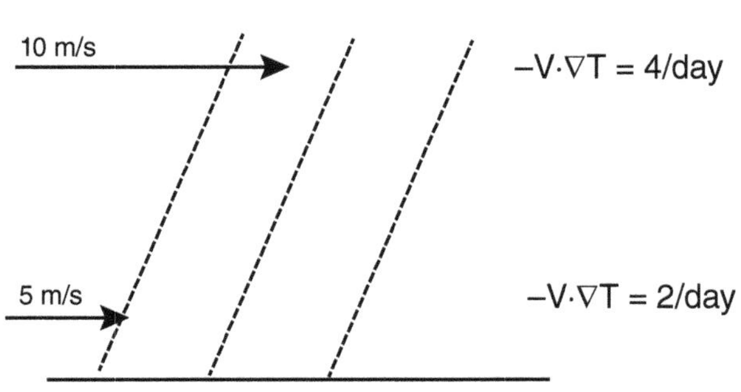

Fig. 10.11 Example of how low-level wind shear leads to differential temperature advection to stabilize the boundary layer.

the increase in stratification is large enough for a given background flow, then the flow can become blocked if the cross-mountain flow is not too strong. As the flow becomes blocked, the winds near the mountains tend to relax or turn mountain parallel (toward low pressure) to the west, with the flow further upstream remaining toward the mountain. This is completely analogous to cold air damming even if the surface air is not that cold. Colder temperatures will allow greater stabilization with even weak warm advection.

This evolution is depicted for a particular case in Fig. 10.12. In this situation, the area is characterized by surface southerly flow initially, as

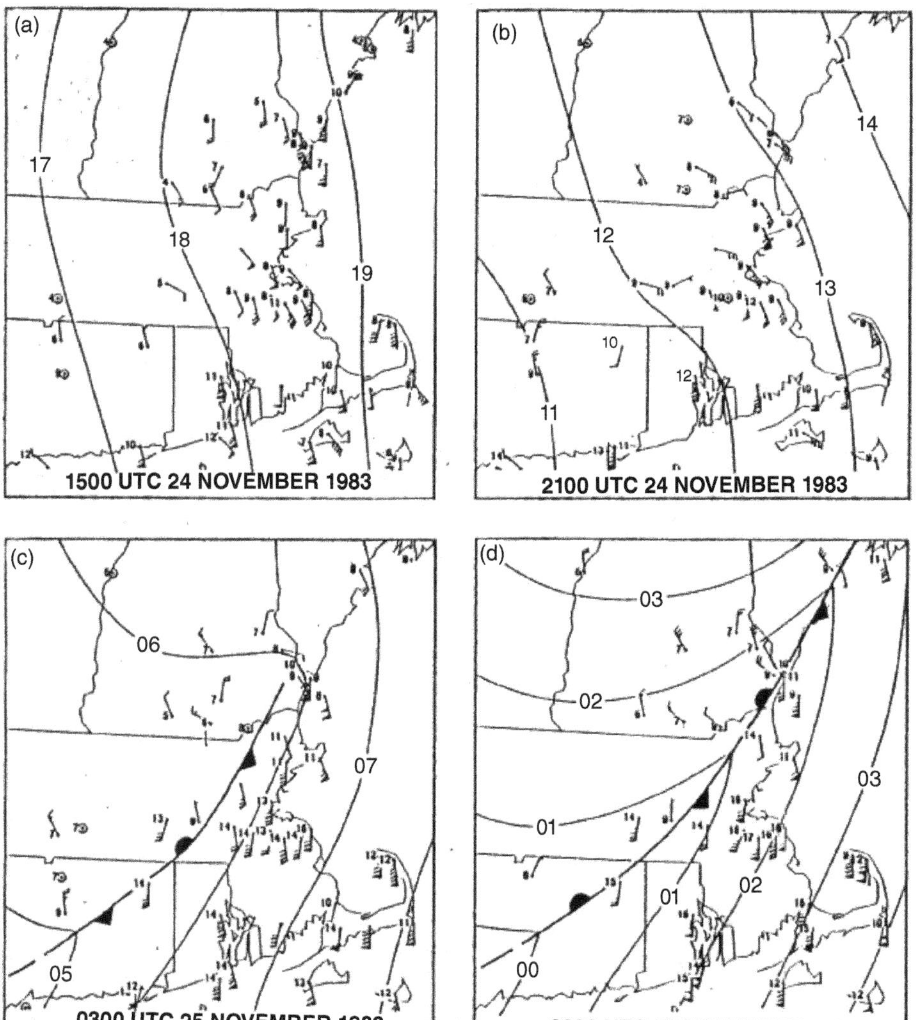

Fig. 10.12

Example synoptic pattern of a type C coastal front event. From Nielsen (1989). Evolution of sea level pressure and winds over 15 hours showing (a) prior to front formation, (b) initial flow blocking, (c) initial front formation, and (d) extension of front northward. ©American Meteorological Society. Used with permission.

shown in the upper left panel of Fig. 10.12. As this flow continues and the air becomes more stable, the upper right panel of Fig. 10.12 shows surface winds over Massachusetts turning easterly in response to the start of flow blocking. As the blocking continues and strengthens, a distinct front is formed as depicted in the lower panels of Fig. 10.12. The important difference between this type of coastal frontogenesis and type A is that the front forms along the base of the topography in the blocked zone and not strictly along the coast. The thermal gradient is increased over time in this case due to warm advection continuing to warm the warm side of the front with little or no temperature change in the blocked region. Nielsen (1989) calls this type C coastal frontogenesis.

10.4 Impacts on Precipitation and Weather

Coastal fronts, as with any other fronts, represent convergent boundaries at the surface that produce upward vertical motion. This lift can contribute to enhanced precipitation along the coastal front. The slope of the baroclinic zone back over the cold air leads to an extended zone of moderate ascent that often produces light precipitation on the cold side of a coastal front. The magnitude of the vertical velocity depends on the shape of the blocked region, which as noted in Chapter 4 can change depending upon the wind shear and stability profiles in the low levels. Depending on the temperature in the cold air, this precipitation can take on a variety of forms from rain to snow or various mixed precipitation types like ice pellets or freezing rain. The details of the precipitation type depend on the vertical structure of the cold air and the overriding warm air. This structure is depicted in Fig. 10.13 which shows the wedge of cold air near the ground

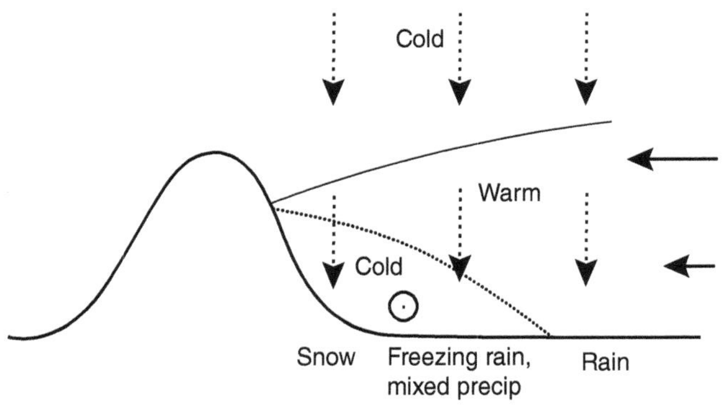

Fig. 10.13 Cross-section through cold air damming and its impact on precipitation type.

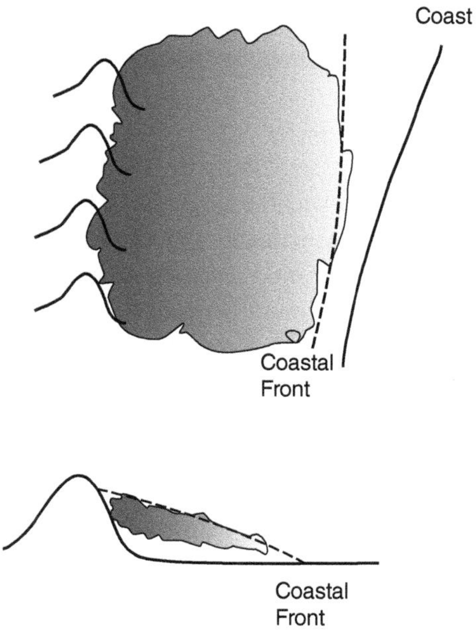

Low-level clouds that occur in the CAD region.

Fig. 10.14

in the blocked region or cold air damming region. This is overlaid with a wedge of warm air (above freezing) and sub-freezing air above this. With precipitation originating in the upper sub freezing layer, it falls into the warm layer and can undergo melting. If the layer near the ground is shallow, the precipitation may not refreeze and be observed as rain. For the deeper cold pool and shallow warm layer, minimal melting occurs, and it falls back into the near-surface cold pool to be observed as snow. In between these extremes, various mixed precipitation scenarios are possible leading to freezing rain, ice pellets, and other hydrometeor types. The details of the vertical structure, the precipitation rate, and the persistence of the warm and cold layers all factor into the detailed evolution of the precipitation and its type.

Low-level clouds often form in the cold air if the boundary layer air is sufficiently moist. The cold air side of a coastal front often behaves like the stratocumulus-filled marine boundary layer off the west coast of North America. The cloud formation is driven by the boundary layer dynamics-causing vertical mixing of near-surface moist air up to its lifted condensation level as shown in Fig. 10.14. The clouds can extend over the entire CAD region as shown in Fig. 10.14. The strong inversion at the top of this layer is formed by the flow of warm air over the cold air that leads to capping of the layer. Precipitation processes in this low-level stratus depend on small-scale dynamics of the boundary layer that can lead to enhanced vertical motion in certain regions.

10.5 Exercises

10.1 Cold air damming is associated with the flow blocking of onshore flow from the east toward topography. Calculate the mountain Rossby radius for the Appalachian mountains assuming a mountain height of 2000 m and a Brunt-Vaisala frequency of $10^{-2}s^{-1}$ .How does this compare to the location of the coast?

10.2 As shown in Chapter 4, the shape of the block depends on the vertical profile of winds and static stability. Explain how a steep block might favor frozen precipitation types and what flow profiles produce a steep block.

10.3 Explain how frictional convergence, due to land–sea friction differences, generates a coastal front when a cold anticyclone moves just offshore to produce geostrophic onshore flow.

11 Coastal Effects on Landfalling Fronts and Cyclones

The interaction of fronts and cyclones with the coast represents a class of coastal phenomena that would not exist in the absence of the synoptic-scale features. While some interaction occurs in the absence of topography, the most pronounced interaction occurs when there is coastal topography. The primary effects and interactions described in this chapter occur as the baroclinic structure of midlatitude fronts and cyclones interact with coastal topographic features. However, topography also impacts tropical cyclones, which will be examined as well. Coastal impacts on tropical cyclones and midlatitude cyclones can also occur without topography and are examined as well. While not a meteorological event, the storm surge produced by cyclones moving toward the coast is a very important impact of coastal or landfalling storms. While there is potential upscale modification by this coastal interaction on the parent cyclone or front, this upscale interaction is not as well understood and beyond the present focus of this book.

11.1 Description of Interactions of Fronts and Cyclones

As considered for various other phenomena, the basic interaction between fronts and cyclones and coastal topography depends upon the Froude number. The Froude number is given as $\frac{U}{Nh}$, where U represents the wind speed of the flow perpendicular to the topography, and N is the Brunt-Viasala frequency that characterizes the flow interacting with the topography. For low Froude numbers (less than 1), the potential for flow blocking exists and may alter the thermal and wind structure around a front or cyclone. For high Froude numbers (greater than 1), the flow goes over the topography which may induce windward ridging and lee troughing effects in the pressure pattern. This in turn alters the low-level flow leading to localized wind maxima or minima. An important aspect of the interaction of fronts and cyclones with coastal features is the temporal variations in the stratification and wind and the type of interaction that it implies. These temporal variations can occur over relatively short time scales as a synoptic-scale cyclone moves across the coast.

While the dynamic impacts on the winds and associated thermal structure is important, the impact on precipitation can be even more significant. There are direct impacts on the precipitation distribution along mountainous coasts due to the mechanically forced ascent by the coastal topography. This orographic precipitation enhancement can be extremely large in some situations. The wind and thermal modifications also feed back on the precipitation structures to alter the precipitation distribution even more. Some or all of the interactions described in the following sections may occur in a single storm.

In the absence of topography, the interaction of cyclones and fronts with the coast depends upon the alteration of low-level structure due to the differing surface exchanges that occur across the coastal boundary. As noted in Chapter 2, the change in surface roughness from ocean to land results in differing surface winds across the boundary that can be convergent or divergent depending on whether the flow is onshore or offshore directed. In addition, the surface exchanges of heat and moisture change across the coastline as well. The impact of these surface changes on midlatitude or tropical cyclones and the types of mesoscale weather features that may be produced depends strongly on where and when they occur within a landfalling storm.

11.2 Wind Effects of Blocking and Barrier Jets

As a midlatitude cyclone approaches the coast and its associated topography, numerous aspects of the cyclone can be altered through interaction with the coast. The type of interaction that is most likely to occur depends upon both the structure of the cyclone and the character of the coast. The situation is depicted in Fig. 11.1, which shows a low-pressure system and associated fronts approaching the West Coast of North America. As the system moves toward the coast, the low-level winds and thermal structure can be altered by the topography in the coastal region. Critical to understanding how this evolution may result in significant coastal interaction with topography is how the Froude number varies in time and along the coast. As a midlatitude cyclone moves toward and across the coast, the coastal environment will experience low-level warm and cold advection, which can have an impact on the stratification that may increase or decrease the Froude number. In addition, the winds around the low also alter the Froude number in the coastal region as the wind strength and direction relative to the topography change as the system propagates across the coast.

Prior to considering the details of the coastal interaction, it is useful to examine the general characteristics of the low-level structure in an idealized midlatitude cyclone. The cyclone can be divided into regions relative to the fronts. Ahead of the warm front, the lower atmosphere and boundary layer tend to be stably stratified and characterized by warm advection associated with the warm front. This warm advection works to increase the stratification over time as the warm front approaches a location. Within the warm sector, between the warm and cold fronts, the atmosphere is relatively warm and stably stratified. This structure is rather uniform through this part of the cyclone. Behind the cold front, the lower atmosphere and boundary layer tend to be weakly stratified or unstable and is characterized by cold advection. The cold advection has a destabilizing effect on the lower atmosphere. The temperature of the underlying ocean has a large impact on the details of the low-level or boundary layer stratification. For example, if the cyclone encounters colder water near the coast, this will have a stabilizing effect due to the low-level cooling that occurs through a downward heat flux. Just the opposite will occur if the water near the coast were warm. The boundary layer will be stabilized due to surface warming by an upward heat flux.

The other aspect of the cyclone that determines its potential interaction with the coast is the wind. Assuming an essentially symmetric storm with uniform winds around the low, the idealized impact of the winds on the Froude number can be assessed. For a storm approaching the west coast, only the flow on the southern part of the storm will produce an onshore component that could be blocked. For a storm approaching the east coast, the north part of the storm produces the onshore flow that can be blocked. The cross-coast component will be small or near zero ahead of the storm, then increase to a maximum at landfall and then decrease after the low passes the coast. The duration and magnitude of the onshore component depend on the storm track as well as the shape of the storm. For example, if a storm tracks directly toward the coast, the onshore flow may get strong relatively quickly, giving a short time period for blocking to occur. If the track is at a 45° angle to the coast, the time until the onshore flow becomes strong is increased by 30%.

To examine the coastal interaction in the pre-warm frontal environment, we can consider the situation shown in Fig. 11.1 where a low-pressure system occurs off the British Columbia coast with a trailing cold and warm front approaching the Oregon and California coast. The low-level isotachs in Fig. 11.1 show relatively strong flow ahead of the warm front and extending down along the cold front. There is also a region of strong winds in the cold sector behind the cold front. The synoptic scale flow above the surface is from the southwest in the warm sector of this approaching cyclone. The 850 hPa analysis (not shown) shows the high potential temperature air in the warm sector with the warm front approaching the coast. The flow just ahead of the warm front is characterized by warm advection as indicated by the arrows in Fig. 11.1. If this warm advection is strongest above the surface, it will tend to stabilize the low levels over time as the front moves toward the coast. This leads to

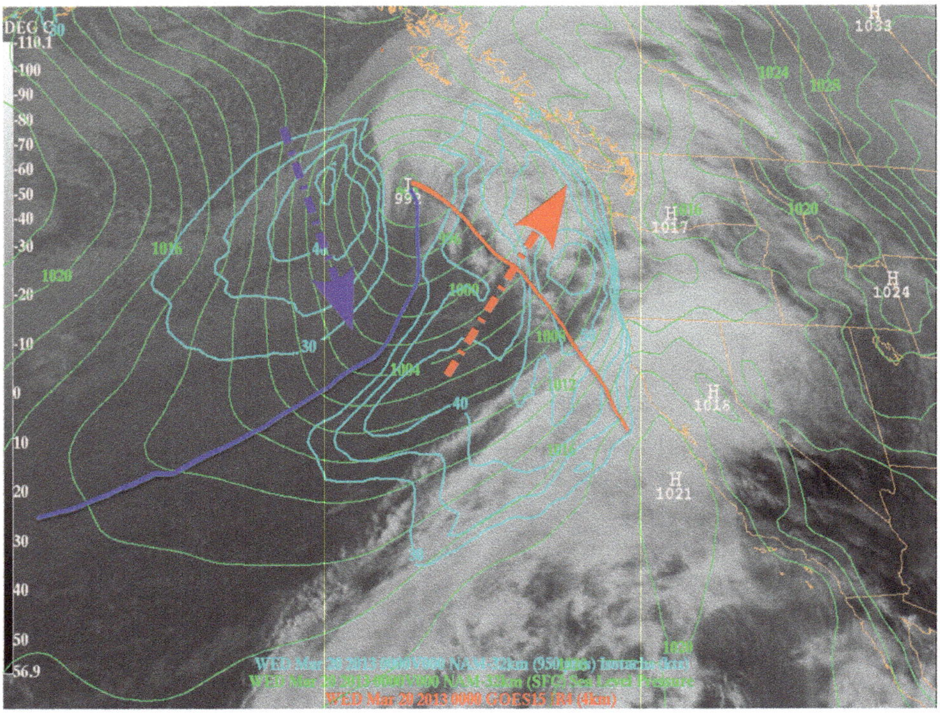

Fig. 11.1 Landfalling cyclone approaching the Oregon and California coast. Green contours are sea-level pressure every 2 hPa, and blue contours are 950 hPa isotachs, every 5kts on infrared satellite image. The warm (red line) and cold front (blue line) are marked. The heavy arrows show the warm (red) and cold (blue) thermal advection near 850 hPa.

the possibility that the low-level flow will become blocked by the coastal topography. Flow blocking will force the flow below the mountaintop level to turn down the pressure gradient and blow in a direction along the mountains toward low pressure. The winds will accelerate in the along-barrier direction to produce a coastal barrier jet.

While strong winds can and do often occur ahead of the warm front in an approaching midlatitude cyclone, the impact of flow blocking on the winds near the coast produces several characteristic structures. To illustrate the impact of flow blocking on the coastal winds and their characteristics, the low-level winds and isotachs are plotted in Fig. 11.2. The figure shows very strong flow along the coast of Oregon just ahead of the approaching cold front. In addition, the flow is enhanced further south along the coast into Northern California relative to what it is further offshore. The wind direction is primarily parallel to the coast and coastal mountains. As covered in Chapter 4, the theory of flow blocking predicts an upstream Rossby radius ($\frac{Nh-u_o}{f}$) that identifies the leading edge of the block. In addition, the flow within the blocked region will be parallel to the barrier. Calculating the Rossby radius for the case shown in Fig. 11.2, we get a Rossby radius of 170 km based on a Brunt-Vaisala Frequency of

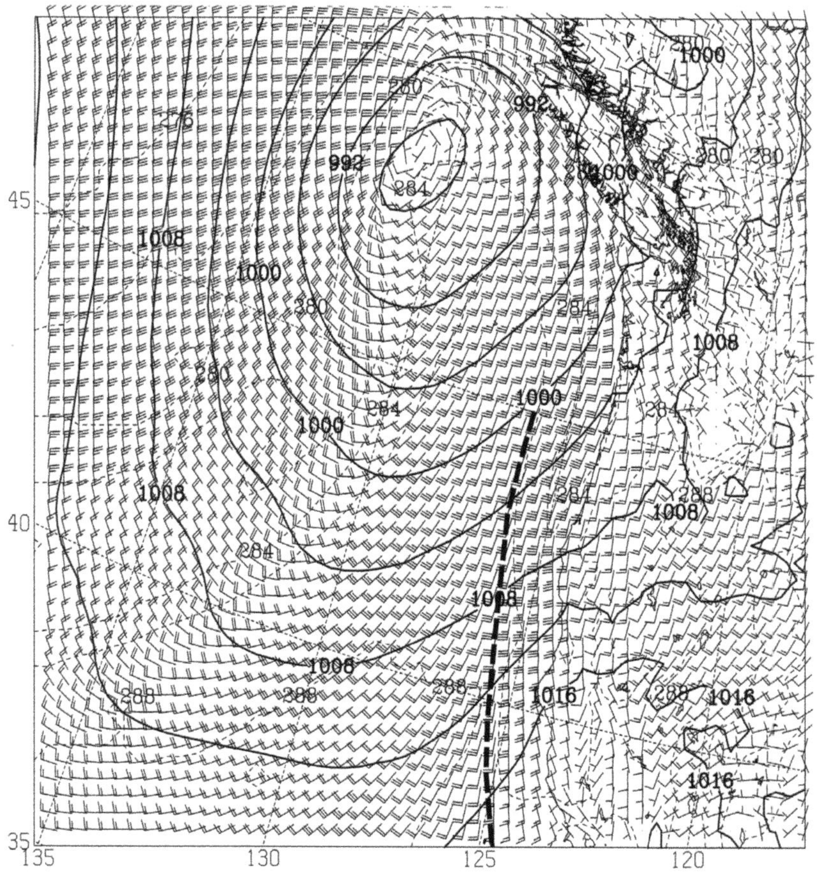

Fig. 11.2 Sea-level pressure (thick solid contours), 950 hPa isentropes (dashed contours), and 10 m winds for a cyclone approaching the West Coast of North America on March 20, 2013 at 0600 UTC. The region of flow that is blocked by the coastal topography is shown by the heavy dashed line.

1.9×10^{-2}, a cross-mountain flow of 10 m/s and a barrier height of 1500 m. This Rossby radius is marked with the heavy dashed line in Fig. 11.2 and corresponds to the location where the winds turn from coast parallel near the coast to more cross-coast further offshore.

The other impact of the flow blocking on the near-shore winds is to cause acceleration along the coastal barrier to create a barrier jet. The strongest winds are observed in Fig. 11.2 right next to the coast. To show the structure of the blocked flow, a cross-section across the Oregon coast shows the presence of a low-level jet below 850 hPa in Fig. 11.3. While the winds are relatively strong above 850 hPa, they increase to over 26 m/s around 950 hPa along the coastal topography. This speed increase is consistent with the thermal wind balance set up by the sloping isentropes and associated horizontal thermal gradient. This is very similar to the summer time jet except that the isentropes slope down offshore to support the southerly flow as opposed to the northerly flow in the summer time jet.

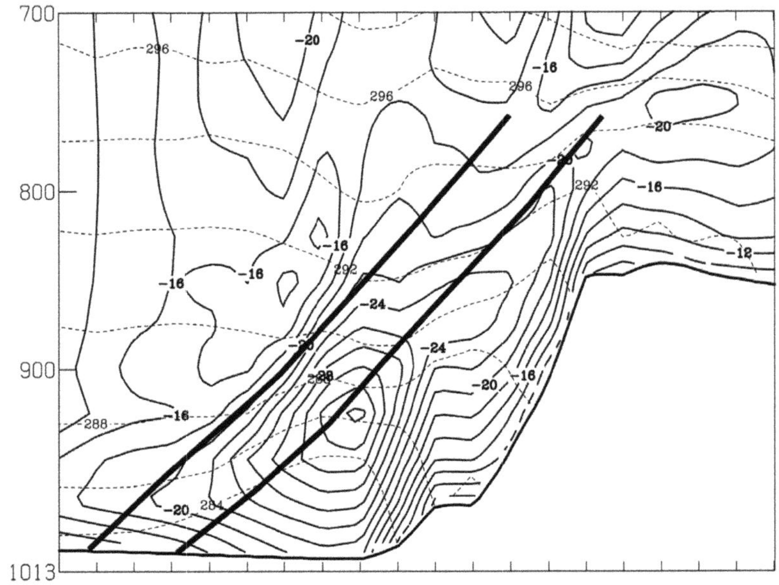

Fig. 11.3 Cross-section of potential temperature and winds perpendicular to the cross-section (along coast flow) as front approaches the coast on March 20, 0600 UTC. The approximate warm frontal zone is marked by the thick solid lines.

If the jet is simply in thermal wind balance with the isentropes associated with the warm front, then it does not necessarily represent a blocked response and associated flow acceleration. The topographic forcing will enhance the flow beyond that implied by thermal wind balance. This can be estimated by calculating the wind shear associated with the thermal gradient and comparing it to the observed shear. Using the thermal wind equation 1.5.11 and estimating the thermal gradient to be 2°/100 km in Fig. 11.3, the wind speed will increase from 16 m/s at 825 hPa to 26 m/s at 950 hPa where the core of the jet is observed. For the jet in Fig. 11.3, the observed winds of 34 m/s are approximately 8 m/s stronger than thermal wind balance implies. This represents the coastal interaction and ageostrophic acceleration that occurs due to the blocking.

The horizontal flow acceleration that occurs in the blocked region can be seen in Fig. 11.2 as the along-shore surface winds increase from 7 m/s to 20 m/s from Northern California to central Oregon. The dynamics of this response can be considered in a manner similar to what was done for gap flows, where the flow acceleration represents a balance between the pressure gradient force and the frictional force. Using equation 9.3.6 from Chapter 9 as given below:

$$u(x) = \left(\frac{-h}{c_d \rho} \frac{\partial P}{\partial x} + \left(u(0)^2 - \frac{-h}{c_d \rho} \frac{\partial P}{\partial x} \right) e^{\frac{-2c_d x}{h}} \right)^{\frac{1}{2}}, \tag{11.2.1}$$

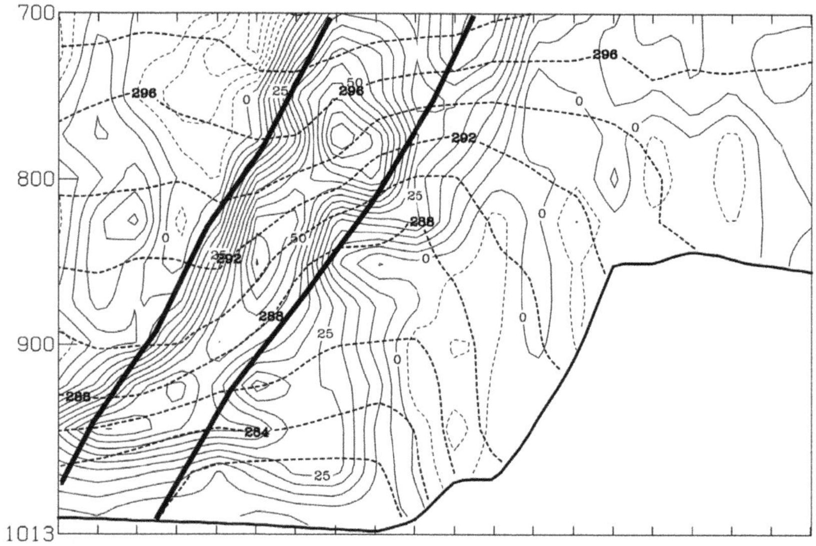

Fig. 11.4

Cross-section of potential temperature and winds perpendicular to the cross-section (along coast flow) as front approaches the coast on March 20, 0300 UTC. The thermal advection is shown by the solid (warm advection)/dashed(cold advection) contours at $5^{o}/day$ intervals. The warm frontal zone is marked by the thick solid lines, which correspond to the strongest warm advection. This warm advection toward the coast acts to increase low-level stability.

we can calculate the wind speed increase that would be expected for the given along-shore pressure gradient. For the situation depicted in Fig. 11.2, the pressure difference between Northern California and central Oregon is about 10 hPa and occurs over a distance of about 400 km. This gives a pressure gradient of 0.0025 pa/m, which we will assume is uniform over the entire distance, as we are interested in just the ending velocity. The height of the friction layer h can be approximated as the level of maximum winds on the cross-section shown in Fig. 11.3. This gives a height of 800 m. While the drag coefficient is not known, an over-water drag coefficient of 3×10^{-3} is probably realistic. Inserting these values into the equation, the predicted winds along the central Oregon coast will be 22 m/s, which is close to what is observed.

To illustrate the stabilization that occurs due to thermal advection in the low-levels near the coast, the cross section in Fig. 11.4 shows the thermal structure and temperature advection ahead of the low as the warm front approaches the coast at 0000 UTC. Warm advection occurs from near the surface offshore to above 850 hPa closer to the coast to set up differential thermal advection. The maximum warm advection is just above the jet, which acts to increase stability in the vicinity of the jet near the coast. The Brunt-Vaisala frequency has increased from $1.0 \times 10^{-2}s^{-1}$ to $1.9 \times 10^{-2}s^{-1}$ in the layer just below the coastal mountains over the 6 hours between 0000 UTC and 0600 UTC. While both static stabilities are sufficient to produce blocking, the weaker stability will only block cross-coast flow of 10 m/s or less. The increase in stability allows the low-level flow to be blocked along the coast as long as the cross-coast wind component is less than 19 m/s. The warm advection shifts the isentropes downward offshore to create the slope and thermal gradient that produces the barrier jet through thermal wind balance. The associated stability increase acts to amplify the along barrier flow due to the flow blocking. Although not shown in the figure, southwesterly flow is occurring near or just above mountaintop level and is relatively unaffected by the coastal mountains. This flow in the 850 – 700 hPa layer is the warm conveyor belt or low-level jet portion of the approaching cold front.

While the increase in static stability may be relatively large, in order for the flow to be blocked, the cross-mountain flow must not get too strong. Typically, the wind speeds will increase as a cyclone approaches the coast even if it not undergoing substantial development. If development is occurring, the wind speed increase may be large. However, the critical aspect for this interaction is the direction of the flow toward the topography. To illustrate how this changes, Fig. 11.5 shows the hypothetical wind evolution at a point along the cross-section. When the low is offshore, winds are southeasterly and the static stability may be relatively low to be unblocked. As the low approaches, warm advection increases and the static stability increases in response. The flow becomes blocked even though the wind speeds may have increased. The cross-barrier portion of the flow is sufficiently small to be blocked and the barrier jet develops into a strong along-barrier flow. While the wind speed overall increases, the cross-mountain flow can stay sufficiently low to allow the flow blocking even as the low gets very close to the coast. During this period of

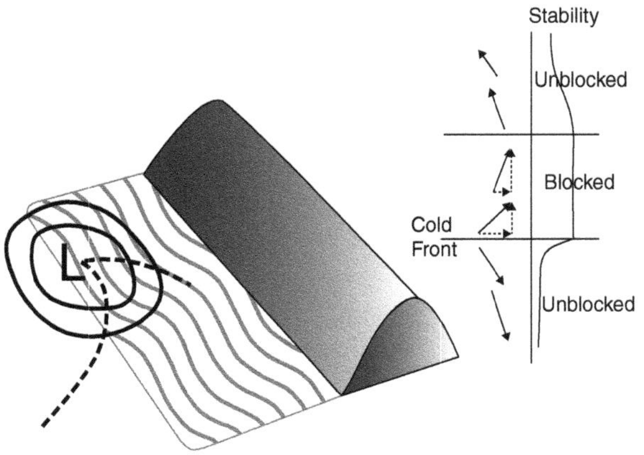

Fig. 11.5 Idealized winds and stability as cyclone approaches the coast. Diagram on the right shows stability and wind evolution as the system approaches. Initially, with low offshore, the winds are south to southeasterly and the static stability is relatively low to produce unblocked conditions. As the low gets closer and warm advection tends to stabilize the low-levels to produce flow blocking. The winds become coast-parallel to give a barrier jet. Then as the cold front passes, the low levels quickly destabilize to give unblocked conditions again.

interaction, the winds along the coast stay coast parallel even though the large-scale pressure gradient is trying to produce southwesterly flow. Then finally as the cold front passes, the low levels quickly destabilize to produce unblocked flow again. The winds then turn westerly to northwesterly in the post-frontal flow. The details of this hypothetical evolution are influenced by the track of the low relative to the topography and how large the cross-mountain flow becomes. If the low tracks more coast parallel, the cross-barrier flow will stay small. If the low approaches more directly, the cross-barrier flow becomes very strong as the flow becomes nearly perpendicular to the topography. In addition to these geometric aspects that produce cross-barrier flow, the intensity of the cyclone also plays a role. Stronger cyclones will generally have stronger winds and hence stronger cross barrier flow. This can prevent flow blocking from being realized.

Pre-frontal flow blocking and the development of enhanced along-coast flow typically occur due to the stratification increases associated with warm advection. This process can be enhanced even more when a gap in the coastal topography allows cold air to flow out from the interior land regions at low levels. Figure 11.6 shows an example of this where strong along-coast winds occur along the coast of Vancouver Island from an east-southeast coast parallel direction. Further offshore to the south, the winds

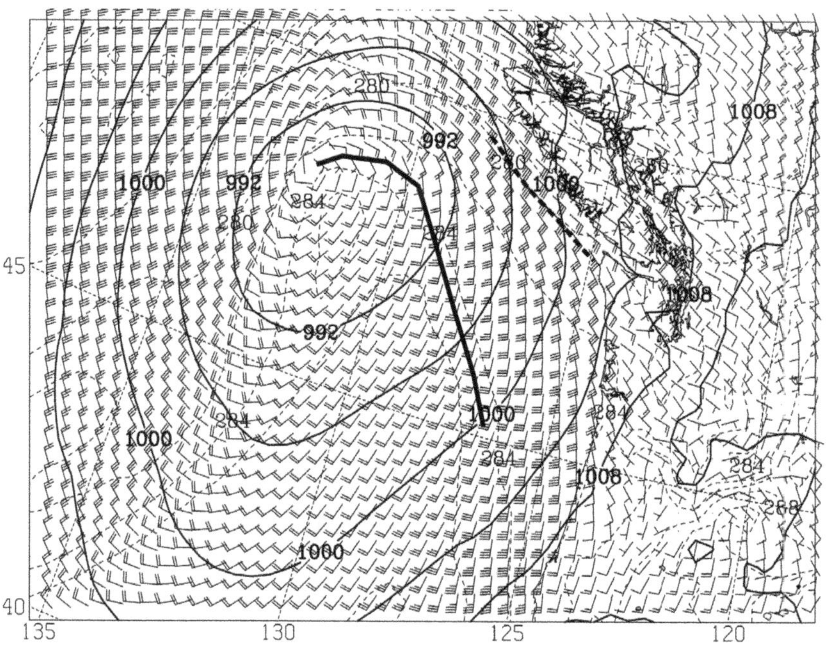

Fig. 11.6

Winds along Vancouver Island as surface low approaches. Solid contours are sea-level pressure every 4 hPa, 1000 hPa isentropes are plotted as dashed contours, and barbs are the winds at 10 m. The low-level winds near the coast of Vancouver Island are within a blocked region marked by the heavy dashed line. The flow within this zone is parallel to the coast. The warm front is marked by a heavy solid line, which occurs further offshore to the south.

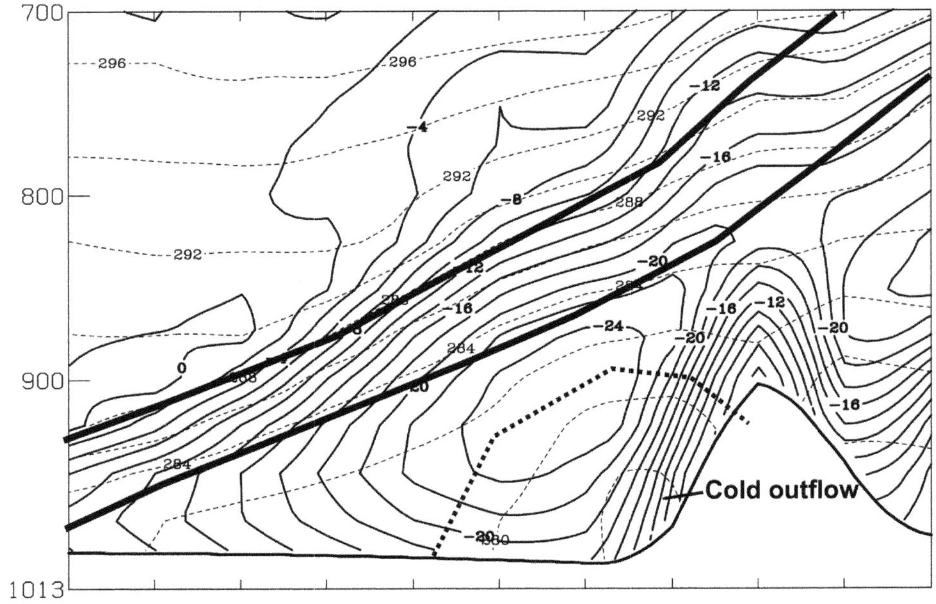

Fig. 11.7 Cross-section through the blocked region upwind of Vancouver Island. Dashed contour lines are potential temperature, which shows a stable layer that slopes up toward the coast of Vancouver Island. The warm frontal zone is indicated by the thick solid lines. A region of colder air near the coast is marked and represents the cold outflow from the Strait of Juan de Fuca. The solid contours are winds perpendicular to the cross section (coast parallel flow) which shows a strong low-level jet near 925 hPa.

become more south-southeasterly and then turn more southwesterly in the warm sector behind the warm front shown in the figure. This wind shift from south-southeasterly to southwesterly is associated with a warm front, as noted in the figure. With the warm front approaching from the south in this case, and we might conclude that the warm advection has established blocking along the coast. A cross-section through the blocked region in Fig. 11.7 shows that there is indeed a sloping stable layer associated with the warm front. The thermal gradient associated with the front produces the low-level jet at about 900 hPa in Fig. 11.7. This stable layer associated with the approaching warm front is still above 1500 m at the coast. However, the low levels below the mountains also have relatively strong static stability near the coast. This pocket of cold air is characterized by coast-parallel flow as seen on the horizontal analysis in Fig. 11.6, which is typical of blocked flow. The cold air at low-levels in this case originates inland over Washington and has been drawn out through the Strait of Juan de Fuca. Overland and Bond (1995) illustrate a similar effect with a much colder outflow that produces a very distinct region of coastal trapping and a barrier jet that are separate from the warm frontal structure. The advection of low-level cold air helps to increase static stability to promote the flow blocking ahead of the approaching front.

Similar effects have been observed along the California coast by Neiman et al. (2004), where a cold outflow from the San Francisco bay region helped to increase stratification and flow blocking ahead of a land-falling front.

Flow blocking in the prefrontal environment impacts the winds and thermal structure as seen in the previous examples. These structural changes also have an important impact on the fronts in the cyclone as they approach the coast. As seen in the case of cold air damming in Chapter 10, the leading edge of blocked region has a thermal gradient associated with it. As a front propagates toward this feature, the low-level front stalls along this boundary due to the flow blocking. The thermal gradient below the topography now becomes aligned with the coastline. Above the topography, the front is able to continue to propagate across the coast with its original orientation. Numerous studies have demonstrated this effect, which can be seen in Fig. 11.2 for the case from March 2013 examined earlier. The thermal gradient is perpendicular to the coast at the surface with strong along-coast flow in the blocked region. Above the surface (not shown), the warm front has moved inland and the coastal region is within the warm sector of this cyclone. This can be seen in the cross section shown in Fig. 11.8 at 0900 UTC. The blocked region has a pronounced barrier jet with isentropes that slope down away from the coast. Above this layer, there is essentially no horizontal thermal gradient as the isentropes are

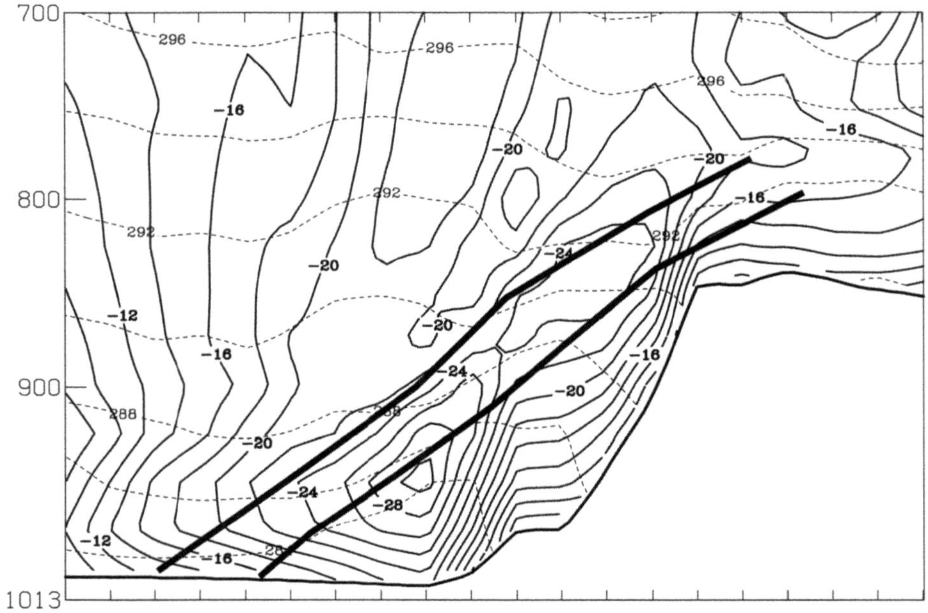

Fig. 11.8 Cross-section of potential temperature and winds perpendicular to the cross-section (along coast flow) as warm front passes the coast at 0900 UTC on March 20. The warm front is marked by the thick solid lines.

mostly flat. This structure remains in place until the cold front passes and the blocked region destabilizes.

11.3 Wind Effects of Flow over Topography

Once a front passes the coast and coastal mountains, the flow blocking is typically eliminated as the static stability of the cold post-frontal air tends to be much lower. This favors flow over the topography that can produce windward ridging and lee troughing type effects. An example of this is shown in Fig. 11.9 from the case, we have examined earlier. As the front passes the coast between 12 and 15Z on March 20, mesoscale pressure ridging and troughing are observed along the coast of Oregon to the south of the low center. The mesoscale ridging and troughing are plotted as perturbations and are associated with westerly flow over the coastal mountains to produce windward ridging over the coastal region and lee troughing inland. The magnitude of the cross-mountain flow in this case is about 20 m/s, and the Brunt-Vaisala frequency is about 2×10^{-3} which yields a pressure response of +/− 1 hPa based on the formula presented in Chapter 8 for flow over isolated topography. This agrees rather well with

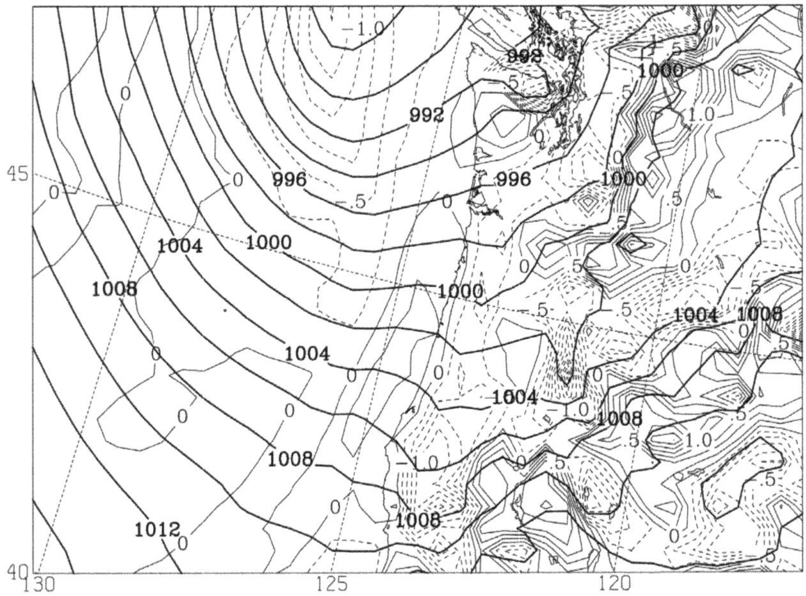

Fig. 11.9 Example of pressure and wind response to topography for unblocked post-frontal flow for March 20, 1200 UTC. The sea-level pressure is the thick solid contours every 2 hPa. The mesoscale perturbation surface pressure is shown by the solid/dashed contours plotted every 0.5 hPa, indicating a pressure ridge (solid contours) just offshore and a pressure trough (dashed contours) inland.

the observed perturbation pressure shown on the plot, where the ridge is +0.5 hPa and the trough is −1 hPa.

This mesoscale windward ridging acts to rotate the pressure gradient force from an along-shore orientation to a more offshore orientation. This turning of the pressure gradient tends to rotate the winds into an along-shore orientation. While this is relatively small for the case in Fig. 11.9, it can be more significant and produce an observable along-coast flow. Figure 11.10 shows a case from October 1999, where the flow along the southern Oregon coast is mostly north–south. The pressure analysis in that region shows a pronounced windward ridge and lee trough that have rotated the pressure gradient to a more cross-coast direction. The impact of the cross mountain flow varies along the coast and is largest in this region, where the cross-coast flow is strongest and the topography is higher as well. Along-coast variations in the windward ridging/lee troughing response can establish localized regions of stronger pressure gradients in the along-coast direction as well. The localized pressure gradient can contribute to increased winds in those regions.

The windward ridge that develops behind the front can also propagate as found by Overland and Bond (1993). As shown in Fig. 11.11, a deep trough and associated low-pressure system are approaching the southeast Alaskan coast. As shown on the surface analysis in Fig. 11.12, the front approaches the coast and crosses the coastal mountains to move inland through the period. During this time, the low is located offshore and a propagates northwestward along the coast. The sea-level pressure analysis shows a windward ridge developing in the post-frontal flow and this feature moves up the coast over time. The figure shows the pressure rise region as it develops once the front moves onshore at 1800 UTC. In a time section analysis, the passage of the front (pressure minimum) and time of rapid pressure rise associated with the mesoscale windward ridge are shown in Fig. 11.13. The pressure rise lags the front by 1–2 hours initially to almost 6 hours as it moves up the coast. The winds plotted in the figure show that the strongest winds occur when the pressure is changing most rapidly. This shows that the strongest winds are an isallobaric response to the mesoscale pressure ridge. The windward ridge propagates more rapidly than the front and is associated with the strongest surface winds. This propagation of this localized high-pressure perturbation could be the result of several possible effects. First, the pressure rise region might be a trapped response that propagates as the flow at the leading edge of the high-pressure perturbation is a cross-coast perturbation flow that is blocked by the coastal topography. The blocked flow must then turn up the coast toward lower pressure forcing an along-coast flow that causes the high to propagate. Second, the along-shore propagation of the pressure rise may represent the rotation of the strongest synoptic-scale winds from southwesterly to southerly and maybe even southeasterly as the upper-level trough moves toward the coast. The localized windward ridging represents the location of the larger pressure response. Based on the analysis presented in

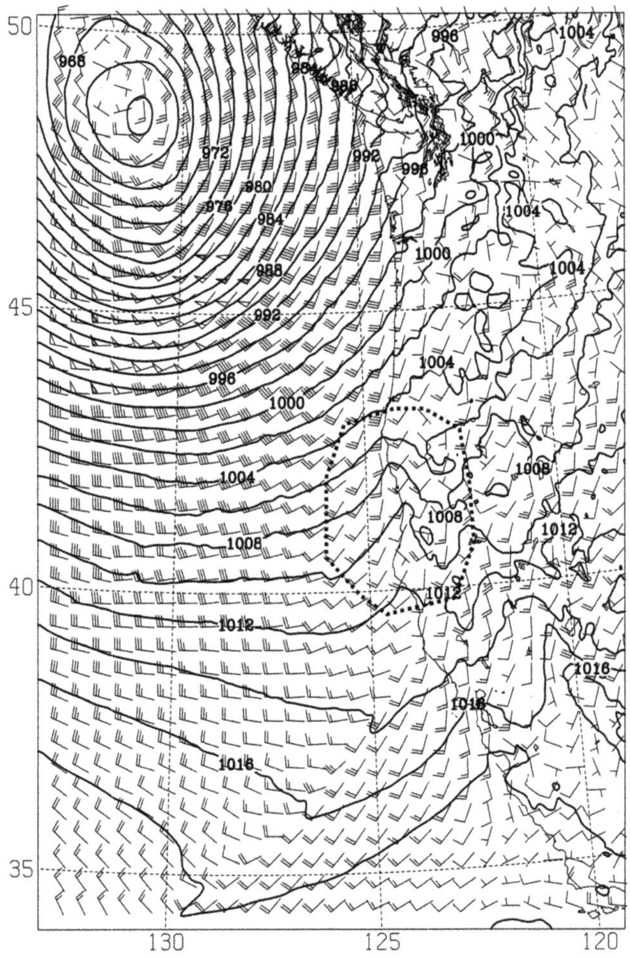

Fig. 11.10 Example of pressure and wind response to topography for unblocked post-frontal flow for October 28 0000 UTC. The sea-level pressure is the thick solid contours every 2 hPa. Wind barbs show the 10 m winds. A localized region of coastal parallel flow is shown, where the winds have rotated due to a mesoscale coastal pressure ridge.

Chapter 7, the response depends on the stability, mountain height, and incident wind speed perpendicular to the mountain. As the mid-level winds rotate, they become more cross-mountain directed along different sections of the coast over time as suggested by the arrows added to Fig. 11.12. This yields the movement or propagation of the windward ridge. Coastal trapping of a post-frontal mesoscale ridge would require sufficient low-level stratification to block the cross-mountain flow that would occur at the leading edge of the disturbance (see discussion of coastal trapping in Chapter 6). While possible to get sufficient stratification in some situations, given that the cross-barrier unblocked flow was the source of the mesoscale

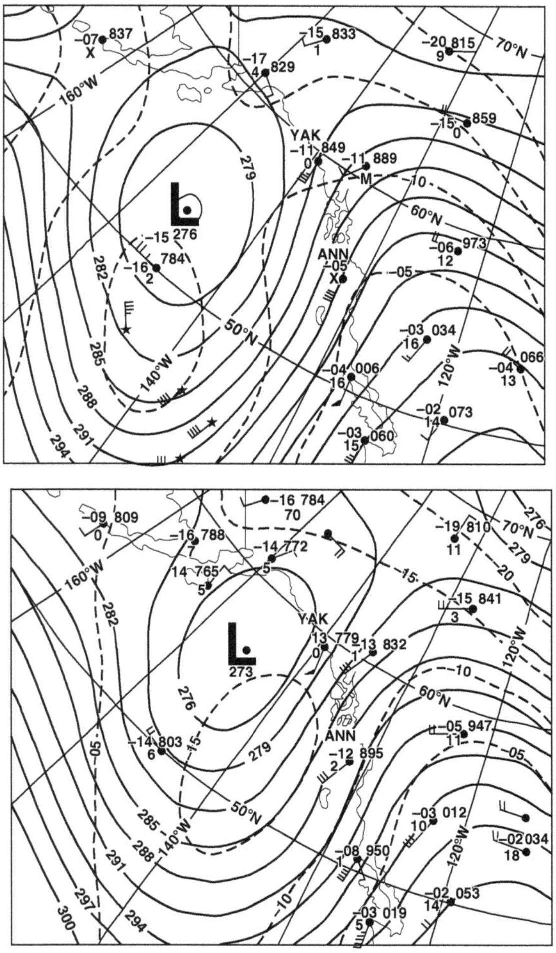

Fig. 11.11

700 hPa analysis of trough approaching the SE Alaska coast. From Overland and Bond (1993). ©American Meteorological Society. Used with permission.

ridge then flow blocking is not expected to trap the response. The mechanism is not entirely clear for the movement of the feature in this case but definitely points to the need to understand the temporal and spatial evolution of the flow and its interaction with the coastal topography.

Thus far, we have focused on the interaction of midlatitude cyclones with coastal topography. Tropical cyclones can also interact with coastal topography in several important ways. The first thing to recognize is that while flow blocking in principle can occur with tropical cyclones, the tropical cyclone environment is generally characterized by moist, potentially unstable flow and relatively strong winds. Consequently, flow blocking is difficult to occur. However, the orographic lift and potential flow over the topography are very significant. In this situation, the flow is forced over the

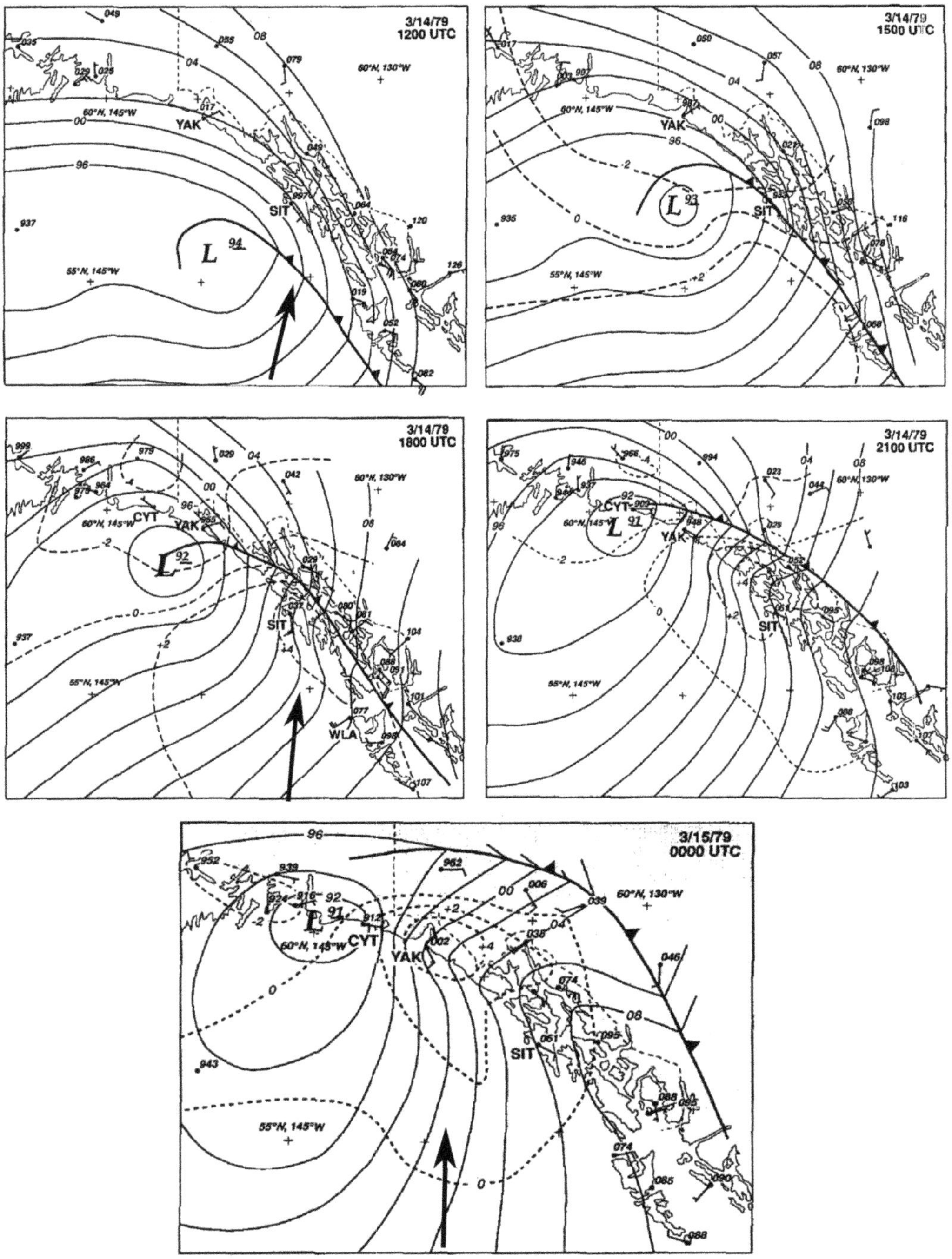

Fig. 11.12 Sea-level pressure analysis and fronts as low approaches coast and front moves onshore. Dashed contours show pressure tendency, which indicates a pressure rise region along the coast that shifts north over time. 700 hPa winds are shown by the thick arrows as inferred from the observations and analysis in Fig. 11.11. From Overland and Bond (1993). ©American Meteorological Society. Used with permission.

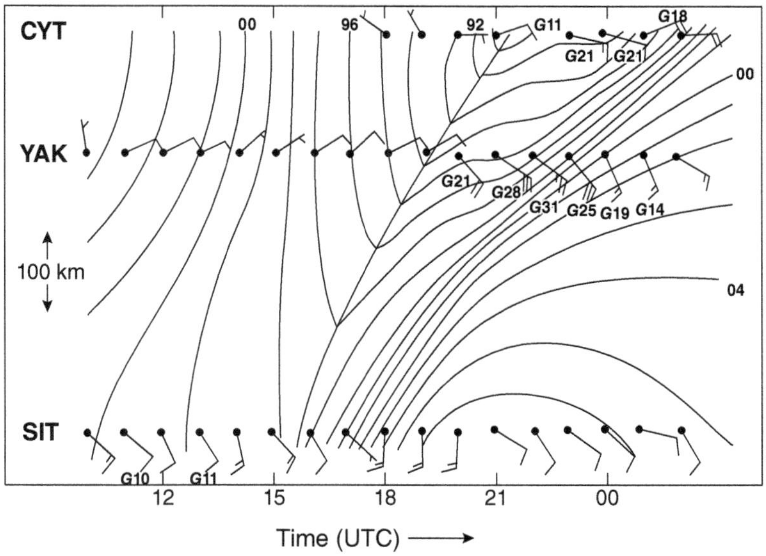

Fig. 11.13 Time series of surface pressure that shows the pressure rise moving faster than the front. From Overland and Bond (1993). ©American Meteorological Society. Used with permission.

terrain following the linear mountain wave dynamics described in Chapter 4. Lee troughing has been observed which tends to reduce precipitation amounts on the lee slope. The windward slope can produce very strong uplift and extreme rainfall.

Another aspect of tropical cyclone interaction with topography is its impact on the tropical cyclone track. Numerous studies have found that the track of a tropical cyclone is impacted upstream of mountainous islands. As highlighted in a study by Huang and Wu (2018), the impact of terrain on the tropical cyclone track comes from two effects. First is the tendency for the large-scale background flow to split around the downstream topography. This results in the track to deflect to the left of the topography as the steering flow splits. This occurs when the tropical cyclone is well upstream of the barrier. As the cyclone gets closer to the topography, the inner-core dynamics interact with the terrain. The flow, while not blocked, intensifies along the barrier to further deflect the tropical cyclone to the left of the barrier. This effect occurs when the inner core is within about 100 km of the terrain.

Up to this point, we have considered the interaction of fronts and cyclones with mountainous coastlines. However, interaction occurs with flat coastlines as well. In this situation, frictional differences may be the most important direct impact. The most direct impact on both tropical and midlatitude cyclones is for the increased friction over land to reduce the circulation and act to dissipate the cyclone strength of a landfalling cyclone. While this impact of friction ultimately influences the entire circulation, the effect of the differential friction across the coast produces some important effects. For onshore flow, the cross-coast frictional difference acts to produce surface convergence. The impact of the friction contributes to the observed tendency for tropical cyclone-induced tornadoes to occur within a few hundred kilometers of the coast and most strongly very near the coast as seen in Schultz and Cecil (2009). The increased friction over land that occurs in the right front quadrant of a landfalling tropical cyclone creates surface convergence and increased vertical motion along the coastal boundary. In addition, the increased friction contributes to enhanced low-level wind shear and horizontal vorticity that can be tilted into the vertical through vertical motion to contribute to tornado formation. While these are not the only factors that contribute to tornadic storms in tropical cyclones, the coastal processes are an important ingredient.

11.4 Precipitation Enhancement

The mesocale low-level structure and winds produced by landfalling fronts and cyclones interacting with coastal topography have an important impact on the coastal weather and its evolution. These modifications to the coastal environment also impact the precipitation amounts and distributions that occur near the coast. The tendency for moist flow over orography to result in enhanced precipitation is a well-known process that extends well beyond just coastal mountains. The modification of clouds and precipitation by flow over topography involves modifications of the dynamic response, thermodynamic response, and cloud microphysical processes through what can be complex interactions. A complete treatment of those interactions is beyond the scope of this discussion, which will focus on the first-order impacts of flow interacting with topography to force uplift and precipitation modifications that can occur as a consequence.

As a moist flow is mechanically forced over mountains, clouds and precipitation develop on the windward slope as a consequence of the adiabatic cooling associated with the vertical motion. This can be either a stable process, such as what might occur in a landfalling frontal system characterized by stable ascent, or as an unstable process, such as what occurs when a potentially unstable layer is lifted to release the instability. Both of these processes will tend to maximize where the strongest ascent is forced. The vertical motion primarily represents the uplift created by flow forced upslope by the topographic feature. The amount of precipitation generally is largest where the largest vertical motion occurs.

The amount of precipitation enhancement produced by the orographic uplift can be approximated by calculating the vertical moisture flux produced by the mechanically forced vertical motion. This can be calculated by taking the dot product between the unmodified, synoptic-scale incident flow and the gradient of the topography. This yields the vertical motion which when coupled with the layer-averaged water vapor mixing ratio in the flow gives a vertical moisture flux. Thus, the precipitation rate can be calculated using the following equation as given in Neiman et al. (2002).

$$P = \overline{\rho q}(v \cdot \nabla z). \tag{11.4.1}$$

The terrain elevation is given by z, v is the incident flow, q is the water vapor mixing ratio, and ρ is the air density. If the layer is essentially saturated, then this vertical moisture flux equates fairly readily to a precipitation rate, with the greatest precipitation rate occurring over the steepest slope where $w_t = v \cdot \nabla z$ is largest. Flow direction and speed variations relative to the topography play a big role in locating the maximum precipitation rate. This process is not a dynamically balanced flow and thus not constrained to a particular scale within the flow. Hence small-scale topographic features can add structure to the precipitation at scales well below the resolution of the winds used to force the ascent. This allows for refining precipitation forecasts based on high-resolution topography that is not fully represented in forecast model wind fields. This topographically forced ascent assumes that the horizontal flow directly interacts with the topography, which occurs when the Froude number is large and the flow is unblocked.

Now the effect of low-level flow blocking changes this process as shown in Fig. 11.14. The blocking tends to shift the location of "orographic" lift away from the mountain slope and toward the coast or even offshore as air is forced up over the block. Both stable and unstable precipitation processes may still be at work, but the location of strongest lift has shifted upstream of the mountain. The result is heavier precipitation at the coast or offshore instead of along the slopes of the topography as shown by Neiman et al. (2002). This shift in precipitation location can have an important impact on determining where flooding might occur in the coastal mountains. This two-dimensional effect of flow blocking is to effectively alter the cross-mountain profile of forced ascent. This provides a first-order impact of the blocking on the precipitation distribution. The shape of the block is important here as it will impact where the strongest uplift may actually occur. As seen in Chapter 4, the low-level wind and stability profile impact the shape of a blocked region. If the wind speed increases upward, as might typically occur, the block tends to be shallow and gradually sloped. If the static stability decreases upward, then the block tends to be deeper and more steeply sloped. In the latter case, the

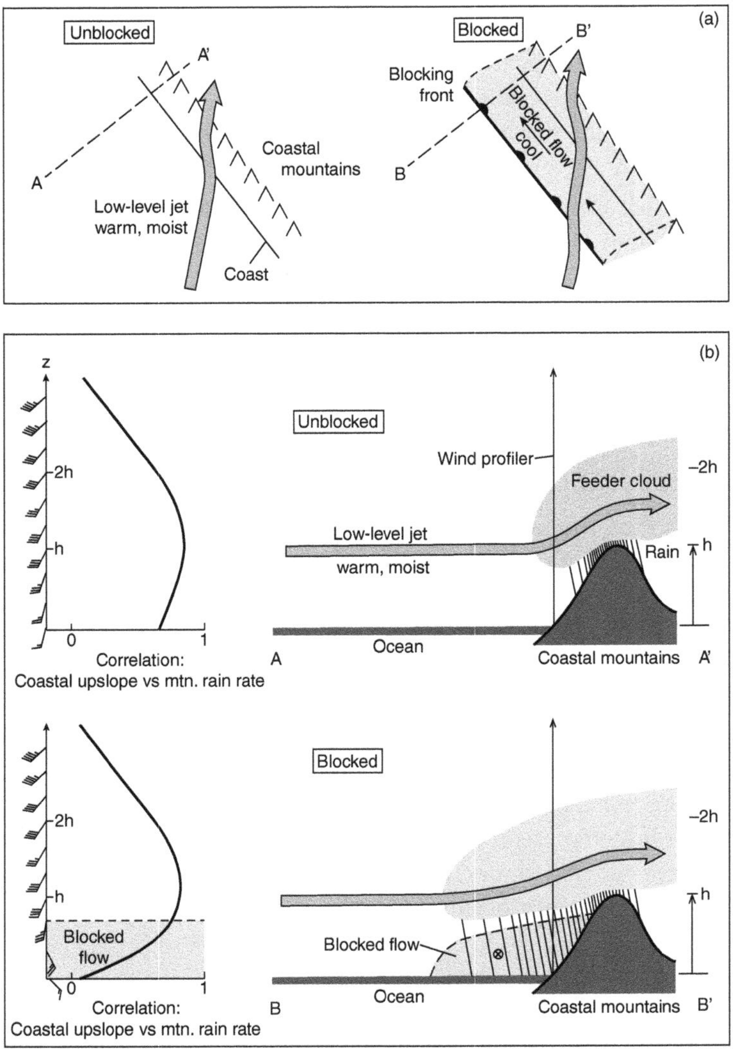

Fig. 11.14 Schematic showing the impact of blocking on orographic precipitation. From Neiman et al. (2002). ©American Meteorological Society. Used with permission.

strongest precipitation is more likely to occur offshore at the leading edge of the block.

For blocked flows, there is a three-dimensional aspect that must be considered as well. In the blocked flow in the low levels, an along-barrier flow results due to the down-gradient ageostrophic flow. However, along-barrier variations in the blocking can occur as seen in Fig. 11.15 or when the edge of the barrier is encountered. These along-barrier variations result in a region where the flow is no longer blocked and low-level convergence occurs between the unblocked incident flow and the along-barrier flow to force stronger ascent in this region. This uplift may produce the strongest precipitation rate at the edge of the topography as illustrated in a study by Rotunno and Ferretti (2001) for a case of flooding in the Alps.

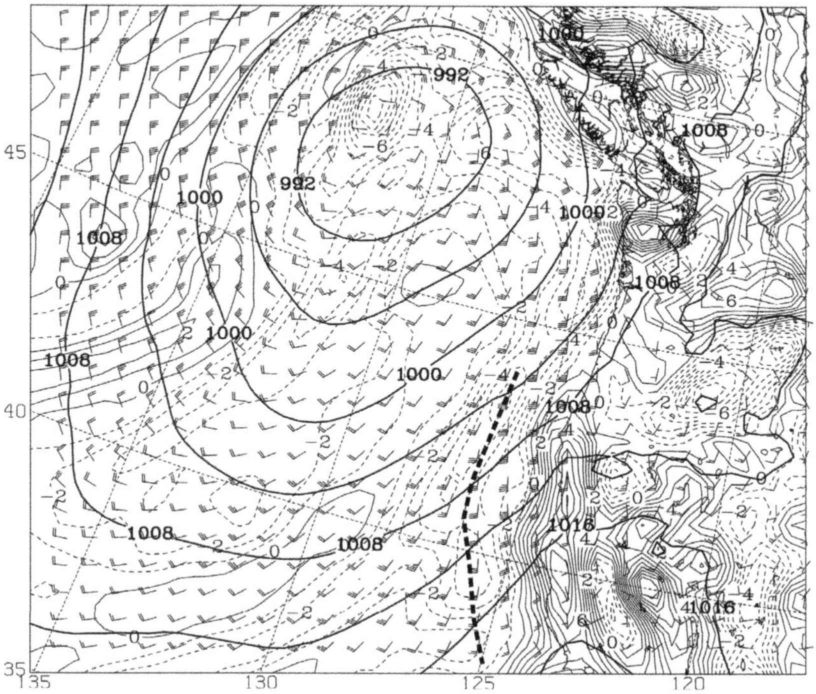

Fig. 11.15 Sea-level pressure (heavy solid contours), surface divergence solid/dashed contours, and 10 m winds for March 20, 0300 UTC. The region of blocked flow and associated Rossby radius are shown by the heavy dashed line, which corresponds to surface convergence. The flow becomes unblocked along the Washington coast, where the convergence shifts inland.

Model resolution certainly influences the ability of the model to capture details in precipitation in coastal topography. While increased resolution will in principle produce the best depiction of these effects, any small-scale errors in flow speed and direction or static stability can greatly alter the details to produce more error than coarse resolution models used to interact with high-resolution topography as noted by Mass et al. (2002).

11.5 Storm Surge

Storm surge is the rise in sea level as a midlatitude or tropical storm approaches the coast. The storm surge occurs due to the combination of winds within a storm, the associated ocean waves, and the coastal ocean bathymetry. Storm surge is the run-up of ocean water onto the coast that occurs when a storm approaches the coast. The surge can happen with both tropical cyclones and midlatitude cyclones. Storm surge is often larger associated with tropical cyclones due to their sustained strong winds and lower pressure. Midlatitude cyclones tend to have weaker winds and higher central pressures which are less favorable for storm surge.

The physical factors that produce storm surge can be divided into three primary processes. The first has to do with the storm pressure and winds producing a circulation in the ocean that impacts the water depth. The second has to do with the generation of ocean surface waves by the storm winds. And the third has to do with the evolution of the ocean waves as they move into the shallow water of the coastal region. The first two processes depend strongly on the characteristics of the storm, including its size, strength, and movement or track. The third process has more to do with the specifics of what happens for a particular coastal region.

Strong winds and low pressure in a cyclone produce a response in the ocean surface water level due to several effects. The first effect is due to the lowered surface pressure in the center of the storm. The lower atmospheric pressure in the storm center has less hydrostatic weight on the underlying ocean. The ocean height will rise by a height equivalent to the pressure reduction in the atmosphere. The amount of sea-level rise due to the pressure difference from outside the storm to the center can be calculated from the hydrostatic equation applied to the ocean as follows:

$$\frac{\partial P}{\partial z} = -g\rho_o. \tag{11.5.1}$$

P is the surface pressure, z is the surface height, g is gravity, and ρ_o is the density of the ocean water. If we approximate this as ΔP representing the change in pressure from the unperturbed environment away from the storm to the center pressure, and then the height change of the sea surface becomes the following:

$$\Delta z = \frac{\Delta P}{-g\rho_o}. \tag{11.5.2}$$

We can evaluate this for a storm with a central pressure of 950 hPa with the pressure away from the storm of 1010 hPa to get $\Delta P = -60$ hPa or -6000 pa. Then using a water density of 1025 kg/m^3 and $g = 9.8$ m/s^2, we get $\Delta z = 0.6$ m. This is a relatively small rise in sea level, and a deeper storm will produce a larger rise. However, even an extremely deep tropical cyclone (less than 900 hPa), will only produce a sea level rise of around 1 m. Consequently, this effect is relatively minor in producing a storm surge. The more significant effect is produced by the winds that result in Ekman-driven transport of water away from the center of a storm. This outward transport sets up a flow of ocean water out ahead of the storm in the direction of movement and downwelling ahead of the storm. The magnitude of this Ekman transport depends upon the strength of the winds as well as the size of a storm. Strong winds and a large storm will produce the largest Ekman transport and greatest downwelling. Away from the coast, the Ekman transport and downwelling are not important, but as the storm gets closer to shore and the water becomes shallow, the downwelling is prevented from occurring and there is a distinct rise in water level. The Ekman transport can be shown to be given by

$$V_E = \frac{\rho_a c_d v^2}{f \rho_w}, \tag{11.5.3}$$

where c_d is the surface drag coefficient, v is the surface wind velocity, f is the Coriolis parameter, and ρ_a and ρ_w are density of air and water, respectively. This gives the expression for the volume of water transported per unit of time over a unit cross-transport distance. The amount of sea-level rise due to the Ekman transport is not as easily calculated and depends upon the depth of the Ekman layer in the ocean and the time scale of the interaction with the shallow water. An approximate estimate of the surge slope can be obtained if we assume a steady state situation with a given Ekman transport occurring in a layer of depth H. In this situation, the shoreward transport in the lower layer is forced up at the coast which then flows back away from the coast to form a wedge of elevated water (highest at the coast). Using these assumptions, the slope of this wedge of water (surge) can be derived to be

$$\frac{\Delta z}{\Delta x} = \frac{\rho_a c_d v^2}{H g \rho_w}, \tag{11.5.4}$$

where Δz is the sea-level rise at the coast and Δx is the offshore distance of the surge. H is the mean water depth of the near-shore ocean and g is gravity. Using some typical values of wind speed 40 m/s, an ocean depth of 50 m and a drag coefficient of 0.01 yields a slope of approximately 4cm per kilometer. If this occurs over a 100 km distance, the sea-level rise would be 4 m. The situation is, of course, not a steady state, and the cyclone is moving toward the coast, so that this estimate may not be particularly accurate. However, it does demonstrate several important aspects of the storm surge. First, the sea-level rise is strongly driven by the magnitude of the winds and stronger winds produce a larger rise. Second, the ocean depth plays a huge role in that shallow water will produce a larger rise than deeper water. Hence, the coastal bathymetry becomes important in determining regions of shallow versus deep water along a coast to produce larger or smaller storm surges.

The second factor that impacts on the sea-level rise as a storm approaches the coast is due to the ocean surface waves generated by the storm winds. The storm-driven waves grow as the winds impart energy into the ocean. The wave growth depends upon the wind speed, the duration, and the fetch associated with the strong winds. The important wave generation region in the storm occurs on the right-hand side of the storm where the winds are in the direction of the storm motion. It is in this region of the storm where the waves will propagate toward the coast. The relative length or fetch of strong winds appears to be relatively small in a storm (maybe a few hundred kilometers), the duration of forcing can be relatively long for a slow-moving tropical cyclone with persistent strong winds. The size of the storm contributes to the wave growth by producing a broader swath of high winds to force waves. This fetch width aids growth by lessening the energy loss due to directional wave dispersion. Depending upon the wave period, the group velocity of the waves can be close to the speed of storm movement to effectively lengthen the fetch, giving greater wave growth. The group velocity for deep water ocean waves is half the wave speed. The wave speed is primarily a function of the wave period and can be approximated by $c = \frac{g}{2\pi}P$ where P is the wave period. This implies that for waves with a $10s$ period, the group velocity is about 8 m/s. So if a storm is moving at that speed, wave growth will be larger. This dynamic fetch allows the waves to grow much larger than the fetch and duration might suggest for a typical storm. Thus storm speed becomes an important factor in aiding wave growth as well. As the waves grow, not only does the wave height increase, but the period becomes longer as well. The wave period becomes an important factor when the waves start to approach the coast and the water becomes shallower. This leads to the third component of the sea-level height rise that is due to shoaling of ocean waves in shallow water and the associated wave runup that happens on the shoreline.

Shoaling occurs when waves propagating in deep water move into shallower water and begin to feel the effects of the sea floor. The critical concept governing this process is that the wave energy is conserved as the waves go from deep to shallow water. The energy propagates at the group velocity ($c_g = c/2$) in deep water, and in shallow water, the wave speed is given by $c = (gh)^{1/2}$ and the group velocity equals the wave speed. Given that the energy is given by the wave height squared, the energy flux is the product of the energy and the group velocity. Requiring that these match between the deep and shallow water, we get the following relationship:

$$1 = \frac{Ec_g}{E_d c_{g_d}} = \frac{H^2}{H_d^2}\frac{c_g}{c_{g_d}}. \tag{11.5.5}$$

The subscript d represents the deep water. Rearranging and solving for the wave height in shallow water, we get the following:

$$H = H_d \times \left(\frac{c_{g_d}}{c_d}\right)^{\frac{1}{2}} = H_d \times K_s, \tag{11.5.6}$$

where K_s is the shoaling coefficient, which can be written as $\frac{\frac{g}{2\pi}P}{(gh)^{\frac{1}{2}}}$. The important aspect to see here is that waves with longer periods will experience greater amounts of shoaling. This implies that storms that have produced the greatest wave growth and longest period waves will have the biggest impact.

Shallow water shoaling becomes an important factor in determining the height of the waves as they break and get forced up the beach. The shoaling depends upon the specific bathymetry of a particular coastline. The shoaling can be calculated using the formula derived above. The effect of the shoaling tends to be largest when the bathymetric slope is gentle as opposed to a steep slope. To further complicate the potential impact of the waves in shallow water, the tendency of the bathymetry to cause energy focusing or dispersion. Submarine canyons focus wave energy and ridges defuse it. Thus the storm surge can vary significantly along a coast due to the specifics of the bathymetry. The effect of the bathymetry implies that coastal bays will typically have wave energy focused in them to cause the largest wave growth and water rise.

While there is no absolutely simple way to calculate the storm surge, the factors that become most important are the depth, size, wind speed, and movement of a storm. These aspects combine to create the sea-level rise, waves, and movement toward the coast. Local bathymetry then alters the response based on how the Ekman transport evolves, and the local wave shoaling and tendency for wave run-up on the beach. Storm surge models bring these factors together which primarily depend on wind speed, storm size, and motion that then interacts with coastal bathymetry.

11.6 Exercises

11.1 The generation of the windward ridge and lee trough in the surface pressure field due to flow across a mountain barrier is a well-known result. The magnitude and structure were covered in Chapter 8. Apply this theory to the flow behind a cold front to determine the magnitude of the perturbation pressure that might occur. Assume the cross mountain flow is $12\,\text{ms}^{-1}$ and the Brunt-Vaisala frequency is $0.005\,\text{s}^{-1}$ and a coastal mountain height of 1500 m.

11.2 Consider the time evolution of flow interacting with topography as a cold front passes a coastal mountain barrier. Assume that the front approaches in a direction such that its orientation is parallel to the mountain barrier. Assume that strong warm advection occurs ahead of the cold front with a maximum near 850 hPa and decreasing warm advection above and below this level. Also, assume that the air behind the cold front is characterized by strong surface mixing that produces a well-mixed marine boundary layer. Given these assumptions:

- Sketch what a representative Skew-T would look like ahead of and behind the cold front.
- Assuming a symmetric structure across the front and surface geostrophic winds of $20\,\text{ms}^{-1}$ from the southwest in the warm air and the northwest in the cold air, calculate the cross-mountain component ahead and behind the cold front.

- Based on the cross-mountain components calculated above and the stratification implied by your profiles in the first part, estimate the Froude number for the pre- and post-cold frontal flows. State what you use numerically to calculate your Froude number. Assume mountains are 1 km high.
- On which side of the front is there potential for flow blocking, and what would make it most favorable (changes to wind speed, stratification, etc.)?
- Given the same numbers, but a different frontal orientation of 45 degrees to the coast vice parallel, describe how the Froude number changes pre- and post-frontally and whether this might be more prone to blocking.

11.3 Explain how landfalling fronts produce a strong along-barrier jet when the warm advection profile promotes strong stratification.

11.4 Calculate the along-barrier wind enhancement that would occur with pre-frontal winds of $20\,\mathrm{ms}^{-1}$ at an angle of 30° to the coast. Assume the static stability is sufficient to block the cross-barrier component and that the flow acceleration happens over a 200 km distance.

Solutions to Exercises

Chapter 1 Exercise Solutions

1.1 Answer

- Surface temperature – This influences the surface heat flux into or out of the atmosphere. The direct impact is limited to the planetary boundary layer and penetrates only to a depth determined by the thermal stratification.
- Surface moisture – This influences the surface moisture flux into or out of the atmosphere. The direct impact is limited to the planetary boundary layer and penetrates only to a depth determined by the thermal stratification.
- Surface roughness – This influences the drag or friction and the associated momentum flux. The direct impact is limited to the planetary boundary layer and penetrates only to a depth determined by the thermal stratification.
- Slope of surface – This influences the air flow near the ground. The direct influence can extend through much of the atmosphere through induced vertically propagating waves. This is a complicated response that depends on numerous factors. At the very least, the change in slope will influence the flow up to the top of the slope.

1.2 Answer

- Time scale of coastal circulation is short compared to the synoptic scale, typically 1 day or less.
- Spatial scale of the coastal circulation is small compared to the geostrophically dominated synoptic-scale flow, typically 10–100 km.
- The large scales force the small scales and not vice versa.
- The depth of coastal circulations is small compared to the troposphere.

1.3 Answer

Initially, the perturbation flow will be oriented in the direction of the perturbation pressure gradient force (toward low pressure). As time goes on, the flow will turn to the right in response to the Coriolis forcing and end up perpendicular to the pressure gradient force after an inertial cycle.

1.4 Answer

- The hydrostatic equation governing the perturbation pressure is as follows:

$$\frac{1}{\rho_0}\frac{\partial p'}{\partial z} = g\frac{\theta'}{\theta_o}$$

- Integrating the hydrostatic equation between the surface and a level z_i yields

$$\int_0^{z_i} \frac{1}{\rho_0}\frac{\partial p'}{\partial z}dz = \int_0^{z_i} g\frac{\theta'}{\theta_o}dz,$$
$$p'(0) = \rho_0 g\frac{\overline{\theta'}}{\theta_o}z,$$

where $\overline{\theta'}$ is the average potential temperature perturbation from the surface up to z_i where the pressure perturbation is assumed to vanish.

- The perturbation pressure calculated using the given parameters is

$$p'(0) = (1.3\,\text{kg/m}^3)(9.8\,\text{m/s}^2)(2°\text{K})/(288\text{K})(1000\,\text{m}) = 88.47\,\text{Pa}.$$

- For $\theta_o = 305$K, the pressure perturbation is 83.54 Pa, and for a temperature rise of 1°C, the pressure perturbation is 44.24 Pa.

1.5 Answer

The thermal wind equation is as follows:

$$\frac{\partial v_g'}{\partial p} = -\gamma\frac{\partial \theta'}{\partial x},$$

and we can evaluate the terms in the 1000 hPa to 850 hPa layer. First, $\gamma = \frac{R}{fP_0}\left(\frac{P_0}{P}\right)^{(1-\frac{R}{c_p})}$ evaluates to 32 using $R = 287$, $c_p = 1004$, $f = 10^{-4}$, and $p_o = 100000$ Pa. Then the thermal wind equation can be approximated as

$$\delta v_g' = -\gamma\frac{\delta\theta'}{\delta x}\delta p,$$

which yields a value of 9.6 m/s given $\delta\theta' = 2°$C, $\delta x = 100000$ m, and $\delta p = 15000$ Pa. If the latitude is changed so that $f = 10^{-5}$, the shear becomes 96 m/s.

Chapter 2 Exercise Solutions

2.1 Answer

Sample skew T-log P charts for the specified conditions:

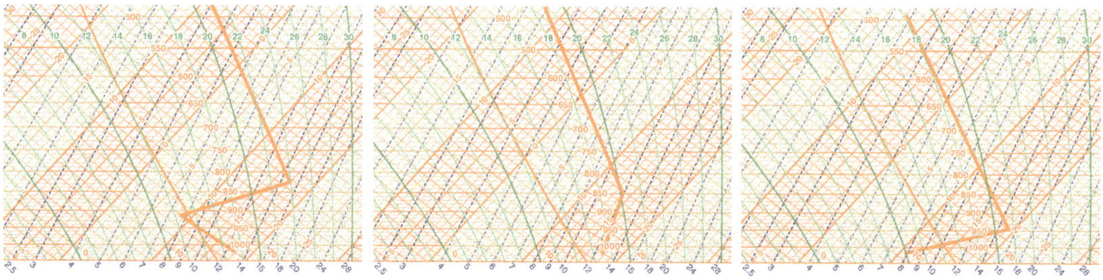

(a) well-mixed marine boundary layer (b) stable marine boundary layer (c) night-times tablel and boundary layer

Fig. A1 Boundary layer temperature profiles under the specified conditions; a) a well-mixed marine boundary layer, b) a stable boundary layer, c) a night-time stable boundary layer.

2.2 Answer

- The well-mixed boundary layer is produced and maintained through buoyant mixing, typically produced by a positive surface heat flux. The mixing can be driven by cooling at the top of the layer to produce negative buoyancy as well. The top of the layer represents where the atmosphere is stably stratified. This can be produced by large-scale subsidence to produce warming and a temperature inversion.
- The stably stratified boundary layer results from either surface cooling or warm advection increases through the layer.
- The nighttime stable boundary layer is produced by surface radiative cooling that makes the near-surface air cool and creates a surface-based inversion. This is maintained until surface heating begins to warm the lowest layers to initiate vertical mixing.

2.3 Answer

- A surface heat flux results in surface warming, which will reduce the stratification in the heated layer to a dry adiabatic lapse rate. The depth of the heated layer depends on the initial lapse rate.
- Differential thermal advection will act to increase the stability for warm advection and decrease the stability for cold advection.

2.4 Answer

- The heat flux can be calculated from the bulk aerodynamic formula,

$$SH = c_p \rho C_h U_{10}(T_s - T_a).$$

Assuming $C_h = 1.4 \times 10^{-3}$ and $c_p = 1004.0$, the surface heat flux is given as

$$SH = (1004.\mathrm{Jkg^{-1}o^{-1}})(1.3\,\mathrm{kgm^{-3}})(1.4 \times 10^{-3})(5\,\mathrm{ms^{-1}})(12°\mathrm{C} - 10°\mathrm{C})$$
$$= 18.27\,\mathrm{Wm^{-2}}.$$

- The latent heat flux can be calculated from the bulk aerodynamic formula,

$$LH = L\rho C_m U_{10}(q_s - q_a)$$

where $L = 2.5 \times 10^6\,\mathrm{Jkg^{-1}}$ and $C_m = 1.15^{-3}$. The specific humidity can be calculated from the dewpoint and pressure (assumed to be 1013 hPa) as follows:

$$e = 6.112 \times e^{\frac{17.67 \times T_d}{(T_d + 243.5)}},$$

to get the vapor pressure e. The specific humidity can then be approximately calculated from

$$q = 0.622 \times \frac{e}{p}$$

which applied to the data given yields $e = 8.72\,\mathrm{hPa}$ and a specific humidity of $q = 5.4\,\mathrm{g/kg}$. Doing the same with the sea surface temperature to get the specific humidity of the ocean surface, we get ($e = 14.02\,\mathrm{hPa}$) and a specific humidity of ($q = 8.61\,\mathrm{g/kg}$). Using these values, the latent heat flux can now be calculated.

$$LH = (2.5 \times 10^6\,\mathrm{Jkg^{-1}})(1.3\,\mathrm{kgm^{-3}})(1.15 \times 10^{-3})(5\,\mathrm{ms^{-1}}),$$
$$((8.61 - 5.4) \times 10^{-3}\,\mathrm{kg/kg}) = 59.99\,\mathrm{Wm^{-2}}.$$

- If the sea surface temperature is increased to 14°C, the sensible heat flux increases to 36.54 $\mathrm{Wm^{-2}}$, and the latent heat flux increases to 82.41 $\mathrm{Wm^{-2}}$.

2.5 Answer

Utilizing the same values as in the text for the wind speed, height of the layer, and Coriolis parameter, increasing the over water drag coefficient from 1.5×10^{-3} to 2.25×10^{-3}, the turning angle increases to 12.7° compared to the 8.5° given in the text.

Chapter 3 Exercise Solutions

3.1 Answer

- The expression governing the evolution of a thermally driven circulation is

$$\frac{dC_a}{dt} = R \ln\left(\frac{P_o}{P_1}\right)(\overline{T}_2 - \overline{T}_1).$$

- Putting the numbers provided into the equation given in the first part of the problem, we get

$$\frac{dC}{dt} = 93.3\,\mathrm{m^2/s^2}$$

for 2°C and

$$\frac{dC}{dt} = 466.4\,\mathrm{m^2/s^2}$$

for 10°C.

- Using $C = Vl$ and first integrating the expression from the previous part over 12 hours, we get $V = 100.7$ m/s for 10°C and $V = 20.1$ m/s for 2°C.
- Both of these results show exceptionally high wind velocities. There are three primary factors that lead to such a large estimate. First is that friction is ignored and it will reduce the flow to a large degree. Second, the heating is assumed to occur over the full depth of the layer when in reality it is large at the surface and goes to zero at the top of the layer. Assuming it were a linear decrease in the vertical, these results would be half the calculated values as the average temperature change is half for a linear profile. Finally, the 12-hour time period is longer than the period of heating that usually occurs. If the thermal rise occurred over 6 hours instead of 12 hours, the wind speed is cut in half.

3.2 Answer

To calculate the temperature rise at the surface, we need the density and volume being warmed. For a unit area of 1 m^2 and a depth of 1 m, the mass being warmed is simply the density times 1 m^3. For water, this is 1000 kg, and for soil, this is 1600 kg assuming a density typical of dry sand. The soil density varies considerably with soil type and moisture content. Given these values for the mass, the temperature rise can be calculated from the expression, $\Delta T = \frac{H \times t}{M \times C}$, where H is the heat flux, C is the heat capacity, M is the mass, and t is the time duration of the heating. Applying this to our two locations, the water temperature rise is 0.45°C and the land temperature rise is 1.4°C.

3.3 Answer

The inertial period is given by $\frac{2\pi}{f}$, which gives a period of 24 hours at 30 N and 13.8 hours at 60 N. The length of daylight in summer and winter depends upon the specific day of year but can be calculated using the following equation:

$$L_d = \left(\frac{24}{\pi}\right) arccos((-tan(d)(tan(\phi))),$$

where d is the solar declination angle and ϕ is the latitude. If we apply this at maximum solar declination (mid-summer (23.5°) and mid-winter (−23.5°)), we get the longest and shortest days. At 30 N, this gives 14 hours in summer and 10 hours in winter, both shorter than the inertial period. At 60 N, this gives 19 hours in summer and 6 hours in winter, where the inertial period is shorter than the day during the summer. The implication is that at 60 N the wind will have a much stronger tendency to adjust geostrophically in the summer.

3.4 Answer

The degree of wind turning from fully ageostrophic to geostrophic is 90°, which presumably will occur over an inertial cycle. While this adjustment or turning depends on details of time-varying pressure distributions and forced vertical circulations, a first-order approximation can be found based on the length of day compared to the inertial cycle for a constant pressure gradient. Doing this, the degree of wind turning is given by the ratio of the length of day to the inertial cycle. Thus at 30 N, the percentage of turning is 58% and 42% in summer and winter, respectively. At 60 N, the percentages are 100% and 43% for summer and winter, respectively.

Chapter 4 Exercise Solutions

4.1 Answer

The upstream distance where blocking effects are no longer important occurs when the flow can maintain or adjust to geostrophic balance. This length scale is given by the Rossby radius defined to be $R = \frac{U}{f}$. The Rossby radius associated with flow blocking is given by $L_R = \frac{(Nh - u_0)}{f}$, where the wind speed is given by the mountain modified flow. For the range of wind speeds from 2 m/s to 10 m/s, the Rossby radius will range from 180 km to 100 km for the higher (more stable) Brunt-Väisälä frequency of 0.02 s^{-1}. For the lower (less stable) Brunt-Väisälä frequency of 0.008 s^{-1}, the Rossby radius ranges from 60 km to −20 km for the wind speeds of 2 m/s and 10 m/s. The negative Rossby radius for the 10 m/s flow implies that the flow is no longer blocked.

4.2 Answer

The air will be cooler near the barrier because as air ascends, the slope it cools adiabatically. This cool region will expand to fill the entire upstream blocked region to form the wedge. Easterly and westerly flows impinging on a mountain barrier can both be blocked and give a very simple vertical structure. The difference occurs when there is a background north-to-south thermal gradient. In this case, the along-barrier flow that forms will advect colder air southward for blocking on the east side of a barrier and warmer air northward on the west side of a barrier. The low-level cold advection acts to increase the static stability helping to maintain the block as the Froude number decreases in the northerly flow case. On the west side of a barrier with low-level warm advection, the warm air tends to weaken the stability and thereby increase the Froude number potentially to the point that the flow will no longer be blocked. Vertical shear in the incident flow acts to change the effective Rossby radius at that level. For stronger flow aloft, the Rossby radius is shortened and for weaker flow below, the Rossby radius is lengthened. This has the effect of flattening the wedge shape in the vertical.

4.3 Answer

The condition for a damped mountain wave response is $k^2 - l^2 > 0$, which implies that the topographic wavenumber k must exceed the wavenumber implied by the Scorer parameter l. The Scorer parameter can be approximated as $l = N/U$ which is the wavenumber of the free atmosphere oscillation or $2\pi\left(\frac{U}{N}\right)$ to get the wavelength. For the parameters given, the implied wavelength is 9425 m or just under 10 km. Thus mountains with a width that is less than 10 km will produce damped mountain waves. Increasing the wind speed to 30 m/s doubles the wavelength to 18.85 km.

4.4 Answer

- First, the lower layer characterized by the conditions given will produce vertically propagating waves as the mountain width of 15 km exceeds the 9.4 km cutoff below which the waves will be damped. To give trapping, the upper layer must produce conditions that produce damped waves. For a 15 km wide mountain, the condition for damping becomes $U > N\left(\frac{w}{2\pi}\right)$ where w is the mountain width. Applying this relationship to this case, any wind exceeding 23.9 m/s will produce damped waves and a trapped wave response.
- The upper layer is damped since the wavelength required to produce damping is given by $2\pi\left(\frac{U}{N}\right)$ which yields 37.7 km or less mountains for the conditions given, well above the 15 km mountains assumed. The lower layer must have a static stability that supports vertical propagation, which can be calculated from $N > \frac{U}{\left(\frac{w}{2\pi}\right)}$ where w is the mountain width. Applied to the 15 km mountain with a 30 m/s flow gives a Brunt-Väisälä frequency greater than 0.013 s^{-1} to give vertically propagating waves.

4.5 Answer

The Froude number is given as $\frac{U}{Nh}$ and blocking radius (Rossby radius) for a mountain modified flow is given as $\frac{Nh-u_o}{f}$. Using these formulas, the following values are found, where the Brunt-Väisälä frequency was calculated as $N^2 = \frac{9.8\ \text{m/s}}{288\text{K}}\frac{12\text{K}}{1000\ \text{m}}$.

Wind Speed (m/s)	Froude Number	Rossby Radius (km)
2	0.099	181
5	0.25	152
10	0.498	102

4.6 Answer

The pressure perturbation can be obtained using the vertical equation of motion and assuming hydrostatic balance. To apply this, we need the mean potential temperature perturbation. The temperature change away from the mountain will be 6.5/km, and over the mountain, it will be the dry adiabatic change of 9.8/km. This implies a temperature difference of 3.3 colder over the mountain at the top of the mountain. Assuming the vertically averaged difference is 1/2 of this value, the perturbation potential temperature difference is 1.65 which can be put into the hydrostatic equation, $p'(0) = \rho_0 g \frac{\overline{\theta'}}{\theta_o}$. Doing so yields a pressure perturbation of 73 Pa.

4.7 Answer

- The mountain width determines whether a geostrophically adjusted response will occur on the high end of the width and whether a damped response will occur on the low end of the width. Between these two extremes, vertically propagating waves or trapped waves can occur, depending on specifics of the background atmospheric structure.
- The wind profile plays a role in determining the mountain wave response. For winds increasing with height, the Scorer parameter can decrease, leading to trapped waves. For winds decreasing with height, either a mean state or self-induced critical level can occur, amplifying the lower level response to produce strong downslope winds.
- The stability profile plays a role in changing the Scorer parameter. If the Scorer parameter decreases with height, this can produce trapping.

4.8 Answer

Downstream low-level static stability that occurs below mountaintop will damp the vertical motion in the induced wave as it tries to penetrate to lower levels. This leads to a decoupling between the surface and wave above, which can limit the impact of downslope flow.

4.9 Answer

The Froude number can be written as $\frac{U}{N(h-z)}$ to account for the remaining mountain height. At z = 0, the Froude number is a minimum, and at z = h, the Froude number goes to infinity and no longer implies any interaction with the mountain. Following this idea for the Rossby radius, the Rossby radius can be written as $\frac{N(h-z)-U}{f}$ assuming a uniform incident flow. This can be rearranged to be $\frac{(Nh-U)-Nz}{f}$ which shows that the blocked region will simply be a linear slope. The slope of the block does not change with wind speed (U) but does change for differing static stabilities (N). If we allow the wind speed to vary in the vertical as well, we can write the Rossby radius as $\frac{(N(h-z)-(U-\Delta U(h-z))}{f}$, where U represents the wind at mountaintop that decreases linearly downward. This can be rearranged to be $\frac{(Nh-U+\Delta Uh)-(N+\Delta U)z}{f}$, which again shows a linear slope to the blocked region. The slope is flatter as the increase in winds through the profile becomes unblocked more quickly.

4.10 Answer

Within the blocked region, the flow becomes subgeostrophic and cannot adjust geostrophically. Thus the flow accelerates toward low pressure in the along-barrier direction. The acceleration can be approximately calculated from the momentum equation:

$$\frac{dv'}{dt} = -\frac{1}{\rho_0}\frac{\partial p'}{\partial y} - fu' + F_y$$

by ignoring the friction and Coriolis forces and approximating the acceleration as $v\frac{\partial v}{\partial y}$. This yields the wind speed as a function of distance to be

$$v(y) = \left(v(0)^2 - \frac{2}{\rho_o}\frac{\partial p}{\partial y}y\right)^{\frac{1}{2}}.$$

The pressure gradient can be obtained from the incident flow speed of 10 m/s using the geostrophic relationship and a Coriolis parameter of 10^{-4}. This gives the along-barrier pressure gradient to be $0.0013\,\mathrm{Pam}^{-1}$. Putting this in the integrated momentum equation, the increase in speed over 100 km would be 16.12 m/s. The strength of the barrier jet is largely determined by the along barrier pressure gradient and the distance over which acceleration can occur. A more complete treatment would include friction which can lead to a balance between the along-barrier acceleration and the friction to get a steady state barrier jet.

4.11 Answer

The condition for undamped vertically propagating mountain waves is that $k^2 - l^2 < 0$, which implies that $k < l = \frac{N}{U}$ ignoring curvature affects in the wind profile. Setting the forcing wave number k to $\frac{2\pi}{w}$, we can see that $w > \frac{2\pi U}{N}$ will give vertically propagating waves. Applying this to the stabilities given, yields that the mountain width must exceed 628 m and 6280 m for Brunt-Väisälä frequencies of 10^{-1} and 10^{-2}, respectively. If the wind speeds were doubled, the mountain widths are doubled to 1256 m and 12566 m.

4.12 Answer

The mean state critical layer acts as a lid through which energy cannot propagate. The energy is forced downward toward the surface to enhance the downslope flow associated with the mountain wave. The lower mean state critical layer will produce the stronger near-surface winds as the energy is trapped in a thin layer.

Chapter 5 Exercise Solutions

5.1 Answer

The maximum winds will occur where the thermal gradient is strongest. The thermal gradient is not uniform in a sea breeze circulation and tends to be strongest at the leading edge of the sea breeze front and secondarily at the coastline. These two locations produce the most acceleration and strongest winds.

5.2 Answer

The two circulations on opposite sides of the lake or channel will develop as the land heats up. The associated circulation will force downward motion over the lake or channel, which gets to be strongest as the two circulations interact. The associated subsidence will produce warming over the water to weaken the cross coast thermal gradient, which in turn weakens the sea breeze circulations.

5.3 Answer

- Based on the assumed stratification, 7°C of temperature increase at the surface will be buoyant and rise to 700 m.
- The gravity current phase speed is given by $C = (g\frac{\delta\theta}{\theta_o}h)^{1/2}$. The mean temperature rise in the 700 m layer is 3.5°C. Assuming the average thermal difference across the sea breeze front is 1.75°C or half the total thermal rise for the layer, and the depth of the layer is 700 m, the phase speed is 6.3 m/s. Assuming that the sea breeze front propagates over a 6-hour period, the inland penetration would be 136 km.
- The maximum wind speed is given by $U_{max} = \frac{kg\overline{\Delta\theta'}Z_i}{(k^2+\Omega^2)\theta_o L}$. Using the same value for k as given in the text, the maximum wind speed will be $U_{max} = \frac{(2\times10^{-5}(9.8)3.5^o(700\,\mathrm{m})}{((2\times10^{-5})^2+(7.292\times10^{-5})^2)300\mathrm{K}(272\,\mathrm{km})} = 1.02\,\mathrm{m/s}$.

5.4 Answer

The speed of advance given an opposing offshore flow is $S = C - V$ where V is the offshore flow speed. Substituting in the gravity wave phase speed and solving for the thermal difference gives $\delta\theta = \frac{(S+V)^2\theta_o}{gh}$. Putting in the values given yields $\delta\theta = 3°\mathrm{C}$.

5.5 Answer

The vertical thermal structure for the sea breeze produces a deep layer of warm air as the heated surface air mixes vertically. The depth of the vertical mixing depends on the stratification. For the land breeze, the near-surface layer cools but does not mix vertically. The depth of the cool layer depends on the time period of cooling such that over longer times, a deeper layer experiences cooling. Given the same near-surface thermal change for either warming, or cooling, the maximum wind speed given by $U_{max} = \frac{kg\overline{\Delta\theta'}Z_i}{(k^2+\Omega^2)\theta_o L}$ will change only due to the depth of the layer. Assuming all other factors are equal, the land breeze speed will be $v_{land} = v_{sea}\frac{z_{land}}{z_{sea}}$ where the z_{land} and z_{sea} represent the depths of the cooling and warming respectively. If the warming goes to 500 m and the cooling only extends to 100 m, the land breeze will be only 20% of the sea breeze speed.

5.6 Answer

The low pressure occurs inland over the Florida peninsula. Given the tendency to adjust geotrophically, the wind on the east coast will turn to the north and on the west coast to the south. The effect of this turning is to reduce the cross-coast component and slow the inland propagation.

5.7 Answer

- The static stability influences the sea breeze in two ways. First, weaker static stability allows for a deeper layer of heating than strong static stability. A deeper warm layer increases the sea breeze intensity. Second, static stability damps vertical motion. Thus strong static stability will result in weaker vertical motions and also act to weaken the overall circulation.
- Sea breeze strength is determined by the vertically averaged thermal difference across the coast. If the vertical depth of the heating is limited, the surface temperature rise can be large as the heat is limited to a shallow layer. If the heating penetrates through a deeper layer, the temperature rise at the surface is less, but the overall circulation is equally strong because a deeper layer is involved.
- The gravity wave phase speed depends on the thermal difference and the depth of the cold layer. A large thermal difference and a deep layer will produce the strongest phase speed and greatest inland propagation.

5.8 Answer

- Using the perturbation hydrostatic equation, $p'(0) = \rho_0 g \frac{\overline{\theta'}}{\theta_o} z$, we can see that the 2° thermal difference gives a pressure difference of 85 Pa.
- The return flow aloft arises because the pressure gradient aloft reverses relative to the surface. If the heated column expands equally upward and downward, then the pressure gradient aloft will be equal and opposite to the surface and the thermal gradient will not produce a surface pressure gradient.
- If the return flow is distributed through a deeper layer, then the opposing pressure gradient at the top of the sea breeze circulation will be weaker and not offset the gradient at the surface as much.

5.9 Answer

The sea breeze fronts must penetrate inland 25 km in order to meet. The average propagation speed is obtained using the gravity current phase velocity which varies with time as the heating varies over time. Based on this, the gravity current phase velocity is $C(t) = \left(g\frac{\frac{(2}{3600)t}}{\theta_o}h\right)^{1/2}$ and the average is assumed to be half the ending phase velocity. The inland movement is given by $\frac{C(t)}{2}t =$ 25000 m. Substituting in the given time-dependent phase speed, we get $\left(\frac{h}{\theta_o}g\frac{(2}{3600)}\right)^{1/2}/2t^{3/2} = 25000$ which can be solved to get t = 8068 s or 2.24 h.

5.10 Answer

The horizontal divergence for the straight coastline occurs in only one direction $\frac{\partial u}{\partial x}$ which can be evaluated as 5 m/s/10 km = 0.0005 s^{-1}. For the bay, additional divergence occurs in the orthogonal direction with opposing 5 m/s flow over the 10 km. The horizontal divergence is now $\frac{\partial u}{\partial x} + \frac{\partial v}{\partial y}$, which evaluates to 0.0015 s^{-1}. Assuming no vertical motion at the top of the layer the vertical velocity at the bottom is simply given by $w = -\nabla \cdot V \times z$. This gives a downward velocity of 0.5 m/s for the straight coast and 1.5 m/s for the bay.

Chapter 6 Exercise Solutions

6.1 Answer

The coastal jet occurs below 850 hPa due to the cross-coast thermal gradient that occurs within the boundary layer below 850 hPa. The vertical shear between the surface and 850 hPa can differ from that implied by the thermal wind equation that relates the vertical shear of the geostrophic flow to the horizontal thermal gradient. However, the thermal wind equation can give a first-order estimate of the vertical shear of the actual flow. Based on this, if the flow were stronger at the top of the layer in the direction of the jet, the resultant jet will be stronger as well.

6.2 Answer

The magnitude of the coastal jet can be calculated using the thermal wind equation $\frac{\partial v_g'}{\partial p} = -\gamma\frac{\partial \theta'}{\partial x}$ as given in 1.5.11b, where $\gamma = \frac{R}{fP_0}\left(\frac{P_0}{P}\right)^{(1-\frac{R}{c_p})}$. If we integrate this equation from 850 hPa down to 1000 hPa, we get

$$v(p_0) = v(p_1) + \gamma\overline{\frac{\partial \theta'}{\partial x}}(p_1 - p_0).$$

Using the values provided and assuming that $f = 10^{-4}$, $\gamma = 32.23$. Putting these into the equation,

$$v(p_0) = 0.0 + 32.23 \times \left(\frac{5}{200000}\right)(85000 - 100000) = -12.09\,\text{m/s}$$

which means for an easterly thermal gradient, the northernly wind will increase by about 12 m/s.

6.3 Answer

The thermal gradient in the 1500 m – 200 m layer is $\frac{12°C}{200000\,m}$. We can apply the thermal wind equation in the same manner as the previous problem to get the wind increase. It is reasonable to assume that the top of the layer at 1500 m is around 850 hPa and the bottom is around 1000 hPa. Inserting into the thermal wind equation, we would get a low-level jet wind speed of 29 m/s. Increasing the inversion strength to 20°C, results in a wind speed of 48.3 m/s.

6.4 Answer

In all cases, the basic ingredients are the same: high pressure offshore with lower pressure inland, coastal topography that can block the low-level flow, and sufficient low-level static stability to prevent cross-barrier flow. The following diagrams show approximate synoptic patterns that result in an along-coast low-level jet.

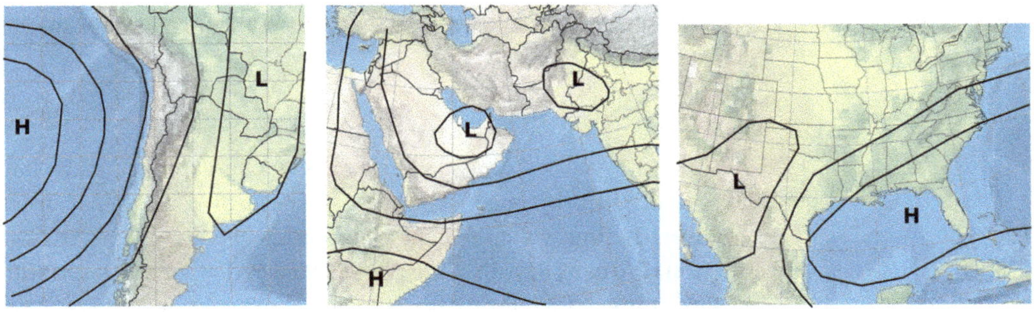

(a) South American Synoptic Pattern (b) Oman Synoptic Pattern (c) Texas Synoptic Pattern

Fig. A2 Synoptic scale sea-level pressure patterns that support a low-level coastal jet for a) the west coast of South America, b) the coast of Oman, and c) the south coast of Texas.

In all cases, the subsidence associated with the high pressure contributes to increased low-level stability and often a definitive inversion.

6.5 Answer

Using the Bernoulli equation given in the text, $\frac{V^2}{2} + g'h = constant$, we can calculate the wind speed that should occur downstream by simply equating the up and downstream conditions.

$$\frac{V_u^2}{2} + g'h_u = \frac{V_d^2}{2} + g'h_d,$$

where $g' = g(\frac{\Delta\theta}{\theta_o})$ and is essentially 0.33 based on the 10°C inversion. Putting in the known values upstream and the known inversion height downstream, we can solve for the downstream wind speed.

$$\frac{(18\,m/s)^2}{2} + 0.33(800\,m) = \frac{V_d^2}{2} + 0.33(300\,m),$$

which gives a downstream wind speed of 25.6 m/s.

6.6 Answer

To get the along-coast acceleration, we need to calculate the along-coast pressure gradient. This is obtained by using the geostrophic wind relationship and solving for the pressure gradient, which gives $\rho_0 f u_g = \frac{\partial p}{\partial y}$. Applied to a 10 m/s flow where the Coriolis parameter is 10^{-4}, the pressure gradient is 0.0013 Pa/m. Taking the sine of 20° and multiplying gives the along-coast component to be 0.00044 Pa/m. The flow acceleration that occurs over a 100 km distance can be obtained using an integrated form of the momentum equation $V_1^2 - V_0^2 = 2\Delta P/\rho$. Assuming $V_0 = 0$, we get a speed increase of 68 m/s.

Chapter 7 Exercise Solutions

7.1 Answer

Coastally trapped wind reversals have been characterized as Kelvin waves and will propagate with a phase speed of $C = (g\frac{\delta\theta}{\theta_o}h)^{\frac{1}{2}}$ for a two-layer shallow water system where $\delta\theta$ is the change across the inversion. We can show that the phase speed is $C = Nh$ for the continuously stratified flow. Using this, the Kelvin wave propagation speed is 12 m/s in this case. The cross-coast length scale is the associated Rossby radius given by $L = \frac{C}{f}$. Using $f = 10^{-4}$ implies a cross-coast length scale of 120 km.

7.2 Answer

In order to produce downslope flow across the coast in the offshore direction, the synoptic-scale pattern must produce offshore-directed flow across-coastal topography. A downslope flow response is produced by exciting mountain waves that focus the energy in the lowest levels. This low-level response is generally enhanced by either a mean state or wave-induced critical layer. The wind profile upstream of the mountains should therefore have the largest cross-barrier flow in the lower levels and decrease with height to produce a wave-induced critical layer or vanish (reverse direction above) to give a mean-state critical layer. The synoptic patterns given in the text show northeasterly flow at low-levels (850 hPa) and westerly flow above (500 hPa) that shows a mean state critical layer will occur somewhere in between. If the warming were simply due to warm advection and not downslope warming, the wind profile only need to produce offshore flow of warm air. Given that warm advection is occurring, the wind profile will exhibit veering (clockwise rotation) in the vertical such that northeasterly low-level flow would shift to easterly and then even southerly above.

7.3 Answer

Using the perturbation hydrostatic equation $p'(0) = \rho_0 g \frac{\overline{\theta'}}{\theta_o} z$, we can calculate the pressure perturbation to be 1.36 hPa. Then the geostrophic low-level flow can be calculated based on a 1.36 hPa pressure change over 200 km. The geostrophic relationship $u_g = \frac{1}{f\rho_0}\frac{\partial p}{\partial y}$ can be used to calculate the flow toward the coast. Putting in the values, gives a geostrophic flow of 5.2 m/s. To determine whether the flow is blocked, we can calculate the Froude number given as $\frac{U}{Nh}$, which is 0.44. This is much less than unity and the flow would be blocked.

7.4 Answer

Stratus surges or coastally trapped wind reversals are the result of a pressure gradient reversal along the coast that is set up by the development of a lee low due to offshore flow. Offshore flow can eliminate the marine boundary layer inversion by warming the boundary layer due to downslope flow. This coastal low-level warming will lower the pressure along the coast to erode the subtropical high and shift it further offshore. Given that the coastal jet occurs along the coastal side of the high, this shift in the high offshore causes the jet to shift offshore as well.

Chapter 8 Exercise Solutions

8.1 Answer

The upstream wind profile needs to support a mountain wave response that favors a strong downslope flow component. A wind profile with stronger flow near the mountain top that decays in the vertical or has a mean state critical layer would favor this response. For the east–west orientation, northerly winds are required. To create a mean state critical layer near 700 hPa, the flow should turn westerly or the speed should drop to zero. The synoptic pattern for eddy events produces low-level northerly flow that turns more westerly aloft, which is consistent with the needed wind profile for downslope winds.

8.2 Answer

Utilizing the geostrophic relationship $u_g = \frac{1}{f\rho_0}\frac{\partial p}{\partial y}$, the 2 hPa perturbation produces a pressure gradient of 0.001 Pa/m which results in a geostrophic wind of 7.7 m/s. Cutting the pressure perturbation in half to 1 hPa reduces the geostrophic flow to 3.85 m/s.

8.3 Answer

The acceleration that will occur due to the perturbation pressure gradient can be calculated from the integrated form of the momentum equation.

$$v(y) = \left(v(0)^2 - \frac{2}{\rho_o} \frac{\partial p}{\partial y} y \right)^{\frac{1}{2}}.$$

Applying this with the values given yields a wind speed increase of 24.8 m/s.

Chapter 9 Exercise Solutions

9.1 Answer

The condition under which ageostrophic down-gradient flow will occur is that the channel width be less than the Rossby radius so that geostrophic adjustment cannot happen. The Rossby radius is given by $R = \frac{U}{f}$ and so the condition for the down-gradient flow is that $W < \frac{U}{f}$. Solving for the wind speed gives $U > Wf$ for the wind speed where an ageotrophic response will be observed. For the channel widths given, this corresponds to winds greater than 3 m/s, and 20 m/s respectively.

9.2 Answer

- Applying the gap wind equation $u(x) = (u(0)^2 - \frac{2\delta P}{\rho})^{\frac{1}{2}}$ without friction, the wind speed is 17.7 m/s. The same calculation can be done to include friction as given by the following equation:

$$u(x) = \left(\frac{-h}{c_d \rho} \frac{\partial P}{\partial x} - \left(u(0)^2 - \frac{-h}{c_d \rho} \frac{\partial P}{\partial x} \right) e^{\frac{-2c_d x}{h}} \right)^{\frac{1}{2}},$$

  which requires a drag coefficient ($c_d = 5 \times 10^{-3}$) over water and using the adjacent mountain heights of 1500 m as the depth of the flow to calculate the wind. Inserting into the equation, the wind speed with friction becomes 14.9 m/s. If the depth of the flow were to be considered as half the mountain height (750 m), then the frictionally modified gap flow becomes 13 m/s.
- The geostrophic flow is given by $u_g = \frac{1}{f\rho_0} \frac{\partial p}{\partial y}$, which can be applied using the along-channel pressure gradient of 200 Pa/100 km to get the geostrophic flow speed. For this pressure gradient, the geostrophic flow is 15.4 m/s.
- Using the equation to calculate the pressure perturbation for flow over isolated topography given in chapter 8, the lee trough and windward ridge pressure perturbation will $p' = -\rho U h N$. Applying this using the geostrophic flow calculated above implies a perturbation pressure response of +/− 3 hPa.
- The cross-channel pressure gradient due to the windward ridging and lee troughing will tend to rotate the pressure gradient into the cross-channel direction. For the magnitude of lee troughing/ windward ridging found in the previous part, the cross-channel

pressure gradient is 0.012 Pa/m and exceeds the along-channel pressure gradient of 0.002 Pa/m. This would support a rather strong along-channel geostrophic flow. Given that the cross-channel flow would not be geostrophic, this effect would be weakened as the trough/ridge pressure perturbation magnitudes would be reduced substantially.

9.3 Answer

The cross-channel flow will go from easterly to southerly and then westerly as the low passes from south to north. With the low to the southwest of the channel, the flow in the channel would be mostly geostrophic as the pressure gradient is mostly cross-channel. As the low moves north, the pressure gradient will become progressively more aligned with the channel. During this time, the flow in the channel will accelerate down the channel. As the low moves further north the pressure gradient will rotate into a more cross-channel direction, which supports a more geostrophic response. To the extent that windward ridging/lee trough occurs with cross-channel flow, this would be strongest when the pressure gradient is along the channel. As this induced pressure gradient is cross-channel, the effect is to weaken the along-channel gradient and delay the initiation of gap winds.

9.4 Answer

Katabatic cross-coast outflow and gap winds will extend offshore until they geostrophically adjust and friction brings them into a three-way balance (geotriptic balance). The Rossby radius, $L_r = \frac{(g'h)^{\frac{1}{2}}}{f}$, gives a reasonable estimate of this distance, where g' is the reduced gravity and typically around 0.33. Depending on the depth of the outflow, the offshore extent can range from 60 km for a shallow 100 m layer to almost 600 km for a deep 1000 m layer.

9.5 Answer

The strongest wind occurs at the gap exit due several effects. First, the flow acceleration continues through the gap until the exit where the flow begins to adjust geostrophically. Second, to the extent that the gap flow is typically cold, the difluent outflow rapidly shallows the cold layer which hydrostatically creates a local pressure gradient at the exit to enhance the acceleration. This effect is strongest when the gap flow is much colder than the surrounding air.

Chapter 10 Exercise Solutions

10.1 Answer

The mountain Rossby radius is given as $\frac{Nh}{f}$ for the maximum upstream distance. For the Appalachian mountains, this is 400 km, which compares quite well with the distance from the mountains to the coast.

10.2 Answer

A steep block would have a more consistently deep cold layer extending out from the topography. A deep cold layer favors frozen precipitation as even if the precipitation is liquid as it falls into the top of the layer, it can often refreeze before reaching the ground for a deep cold layer.

10.3 Answer

Coastal frontogenesis requires convergence at the coast to concentrate the thermal gradient. Frictional convergence occurs at the coast whenever there is cross-coast onshore flow. The winds over land turn more strongly toward low pressure than the winds over the water with less frictional drag. The frictional convergence might be weak initially as the geostrophic flow is only slightly onshore; however, the effect gets stronger as the geostrophic flow becomes progressively more onshore.

Chapter 11 Exercise Solutions

11.1 Answer

Using the formula for the perturbation pressure response to flow over topography $p' = -\rho UhN$ to get the amplitude of the perturbation, the pressure ridge should be 1.17 hPa.

11.2 Answer

- Representative Skew-T's ahead of and behind the cold front are shown.
- The cross-mountain wind component will be $u = V cos(45)$ based on the winds from the southwest or northwest with the front parallel to the barrier. This gives a cross-barrier wind of 14.1 m/s from the west for both the pre- and post-frontal winds.
- The Froude number is $\frac{u}{Nh}$. If we assume a static stability given by a Brunt-Väisälä frequency of 0.02 s^{-1} for the pre-frontal flow, then the Froude number will be 0.7. For the post-frontal flow, the static stability might be 0.005 s^{-1}, and the Froude number will be 2.82.
- The pre-frontal flow is potentially blocked (based on the assumed Brunt-Väisälä frequency) and the post-frontal flow is not blocked. Increasing the wind speed would reduce the blocking, while increasing static stability would increase the potential.
- If the front is rotated 45°, the pre-frontal flow is directly toward the coast and the post-frontal flow is along the coast. The

pre-frontal flow would not be blocked as the cross barrier flow increases from 14.1 to 20 m/s, giving a Froude number of 1. The post-frontal flow would have no cross-coast component and would not be blocked.

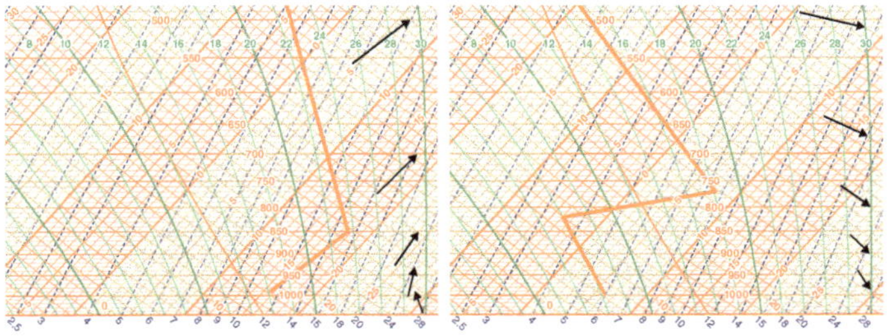

(a) pre-frontal sounding (b) post-frontal sounding

Example temperature and wind profiles a) ahead of and b) behind a cold front plotted on a Skew-T diagram. Fig. A3

11.3 Answer

The warm advection profile with stronger warm air advection above mountaintop level acts to increase static stability to force the flow to be blocked. Once the flow becomes blocked, the flow must accelerate down the pressure gradient in the along-barrier direction. This sets up a barrier jet.

11.4 Answer

To calculate the along-barrier flow acceleration, we need the along-barrier pressure gradient. The pressure gradient can be calculated using the cross-barrier flow component, which is $Vsin(30)$ and equal to 10 m/s for the 20 m/s total flow. Using the geostrophic wind relationship $u_g = \frac{1}{f\rho_0}\frac{\partial p}{\partial y}$ and solving for the pressure gradient, the along-barrier pressure gradient is 0.0013 Pa/m based on the geostrophic flow of 10 m/s. Using this pressure gradient, the flow increase is $v(y) = (v(0)^2 - \frac{2}{\rho_o}\frac{\partial p}{\partial y}y)^{\frac{1}{2}}$. Assuming no initial flow $v(0) = 0$, the result speed increase is 32.24 m/s over the 200 km distance.

Bibliography

Arritt, Raymond W. (1993). "Effects of the large-scale flow on characteristic features of the sea breeze." In: *Journal of Applied Meteorology and Climatology* 32.1, pp. 116–125.

Bond, Nicholas, Clifford F. Mass, and James E. Overland (1996). "Coastally trapped wind reversals along the United States west coast during the warm season. Part I: Climatology and temporal evolution." In: *Monthly Weather Review* 124.3, pp. 430–445.

Colle, Brian A. and Clifford F. Mass (2000). "High-resolution observations and numerical simulations of easterly gap flow through the Strait of Juan de Fuca on 9–10 December 1995." In: *Monthly Weather Review* 128.7, pp. 2398–2422.

Cross, Patrick S. (2003). *The California Coastal Jet: Synoptic Controls and Topographically Induced Mesoscale Structure*. Naval Postgraduate School.

Dorman, Clive E. (1985). "Evidence of Kelvin waves in California's marine layer and related eddy generation." In: *Monthly Weather Review* 113.5, pp. 827–839.

Dorman, Clive E., David P. Rogers, Wendell A. Nuss, and William T. Thompson (1999). "Adjustment of the summer marine boundary layer around Point Sur, California." In: *Monthly Weather Review* 127.9, pp. 2143–2159.

Doyle, James D. and Melvyn A. Shapiro (1999). "Flow response to large-scale topography: The Greenland tip jet." In: *Tellus A: Dynamic Meteorology and Oceanography* 51.5, pp. 728–748.

Dunn, Lawrence (1987). "Cold air damming by the Front Range of the Colorado Rockies and its relationship to locally heavy snows." In: *Weather and Forecasting* 2.3, pp. 177–189.

Durran, Dale R. (1986). "Mountain waves." In: *Mesoscale Meteorology and Forecasting*. Springer, pp. 472–492.

Emanual, Kerry (1984). "Fronts and Frontogenesis: Other Types of Fronts." In: *Dynamics of Mesoscale Weather Systems*. National Center for Atmospheric Research.

Estoque, Mariano A. (1962). "The sea breeze as a function of the prevailing synoptic situation." In: *Journal of the Atmospheric Sciences* 19.3, pp. 244–250.

Ferber, Garth K. and Clifford F. Mass (1990). "Surface pressure perturbations produced by an isolated mesoscale topographic barrier. Part II: Influence on regional circulations." In: *Monthly Weather Review* 118.12, pp. 2597–2606.

Gangoiti, Gotzon, Ana Rodriguez-Garcia, Estibaliz Saez de Camara, Eduardo Torre-Pascual, Maria Carmen Gomez, Maite de Blas, Jose Antonio Garcia, Estibaliz Garcia-Ruiz, Inaki ZUazo, Veronica Valdenebro, and John Iza. (2023). "Galernas: A history of coastally trapped disturbances (2003–2020) with hidden frontogenesis in the Bay of Biscay." In: *Atmospheric Research* 281, p. 106493.

Holton, James R. (2004). *An Introduction to Dynamic Meteorology. 4th edn*. Academic Press, Inc.

Hsu, Shih-Ang (1970). "Coastal air-circulation system: Observations and empirical model." In: *Monthly Weather Review* 98.7, pp. 487–509.

Hsu, Shih-Ang (2013). *Coastal Meteorology*. Elsevier.

Huang, Kuan-Chieh and Chun-Chieh Wu (2018). "The impact of idealized terrain on upstream tropical cyclone track." In: *Journal of the Atmospheric Sciences* 75.11, pp. 3887–3910.

Macklin, S. Allen, Gary M. Lackmann, and Judith Gray (1988). "Offshore-directed winds in the vicinity of Prince William Sound, Alaska." In: *Monthly Weather Review* 116.6, pp. 1289–1301.

Mahrt, Larry (1982). "Momentum balance of gravity flows." In: *Journal of the Atmospheric Sciences* 39.12, pp. 2701–2711.

Mass, Clifford F. and Mark D. Albright (1987). "Coastal southerlies and alongshore surges of the west coast of North America: Evidence of mesoscale topographically trapped response to synoptic forcing." In: *Monthly Weather Review* 115.8, pp. 1707–1738.

(1989). "Origin of the Catalina eddy." In: *Monthly Weather Review* 117.11, pp. 2406–2436.

Mass, Clifford F. and Nicholas A. Bond (1996). "Coastally trapped wind reversals along the United States west coast during the warm season. Part II: Synoptic evolution." In: *Monthly Weather Review* 124.3, pp. 446–461.

Mass, Clifford F., David Ovens, Ken Westrick, and Brian Colle (2002). "Does increasing horizontal resolution produce more skillful forecasts? The results of two years of real-time numerical weather prediction over the Pacific Northwest." In: *Bulletin of the American Meteorological Society* 83.3, pp. 407–430.

Moore, G. W. Kent and Ian A. Renfrew (2005). "Tip jets and barrier winds: A QuikSCAT climatology of high wind speed events around Greenland." In: *Journal of Climate* 18.18, pp. 3713–3725.

National Research Council (NRC) (1992). *Coastal Meteorology: A Review of the State of the Science*. National Academies Press.

Neiburger, Morris (1961). *Studies of the Structure of the Atmosphere over the Eastern Pacific Ocean in Summer*. Vol. 1. University of California Press.

Neiman, Paul J., P. Ola G. Persson, F. Martin Ralph, David P. Jorgensen, Allen B. White, and David E. Kingsmill (2004). "Modification of fronts and precipitation by coastal blocking during an intense landfalling winter storm in southern California: Observations during CALJET." In: *Monthly Weather Review* 132.1, pp. 242–273.

Neiman, Paul J., F. Martin Ralph, Allen B. White, David E. Kingsmill, and P. Ola G. Persson (2002). "The statistical relationship between upslope flow and rainfall in California's coastal mountains: Observations during CALJET." In: *Monthly Weather Review* 130.6, pp. 1468–1492.

Nielsen, John W. (1989). "The formation of New England coastal fronts." In: *Monthly Weather Review* 117.7, pp. 1380–1401.

Nielsen, John W. and Peter P. Neilley (1990). "The vertical structure of New England coastal fronts." In: *Monthly Weather Review* 118.9, pp. 1793–1807.

Nuss, Wendell A. (2007). "Synoptic-scale structure and the character of coastally trapped wind reversals." In: *Monthly Weather Review* 135.1, pp. 60–81.

Overland, James E. and Nicholas A. Bond (1993). "The influence of coastal orography: The Yakutat storm." In: *Monthly Weather Review* 121.5, pp. 1388–1397.

Overland, James E. and Nicholas A. Bond (1995). "Observations and scale analysis of coastal wind jets." In: *Monthly Weather Review* 123.10, pp. 2934–2941.

Overland, James E. and Bernard A. Walter Jr. (1981). "Gap winds in the Strait of Juan de Fuca." In: *Monthly Weather Review* 109.10, pp. 2221–2233.

Parish, Thomas R. and David H. Bromwich (2007). "Reexamination of the near-surface airflow over the Antarctic continent and implications on atmospheric circulations at high southern latitudes." In: *Monthly Weather Review* 135.5, pp. 1961–1973.

Pierrehumbert, Raymond T. (1986). "Lee cyclogenesis." In: *Mesoscale Meteorology and Forecasting*. Springer, pp. 493–515.

Ralph, F. M. et al. (1998). "Observations and analysis of the 10–11 June 1994 coastally trapped disturbance." In: *Monthly Weather Review* 126.9, pp. 2435–2465.

Raman, S., Neeraja C. Reddy, and Devdutta S. Niyogi (1998). "Mesoscale analysis of a Carolina coastal front." In: *Boundary-Layer Meteorology* 86.1, pp. 125–145.

Reed, Richard J. (1980). "Destructive winds caused by an orographically induced mesoscale cyclone." In: *Bulletin of the American Meteorological Society* 61.11, pp. 1346–1355.

Rotunno, Richard and Rossella Ferretti (2001). "Mechanisms of intense Alpine rainfall." In: *Journal of the Atmospheric Sciences* 58.13, pp. 1732–1749.

Schultz, Lori A. and Daniel J. Cecil (2009). "Tropical cyclone tornadoes, 1950–2007." In: *Monthly Weather Review* 137.10, pp. 3471–3484.

Scorer, Richard S. (1949). "Theory of waves in the lee of mountains." In: *Quarterly Journal of the Royal Meteorological Society* 75.323, pp. 41–56.

Skamarock, William C., Richard Rotunno, and Joseph B. Klemp (1999). "Models of coastally trapped disturbances." In: *Journal of the Atmospheric Sciences* 56.19, pp. 3349–3365.

Skamarock, William C., Richard Rotunno, and Joseph B. Klemp (2002). "Catalina eddies and coastally trapped disturbances." In: *Journal of the Atmospheric Sciences* 59.14, pp. 2270–2278.

Smith, Ronald B. (1980). "Linear theory of stratified hydrostatic flow past an isolated mountain." In: *Tellus* 32.4, pp. 348–364.

Smith, Ronald B. (1982). "Synoptic observations and theory of orographically disturbed wind and pressure." In: *Journal of the Atmospheric Sciences* 39.1, pp. 60–70.

Winant, C. D., C. E. Dorman, C. A. Friehe, and R. C. Beardsley (1988). "The marine layer off northern California: An example of supercritical channel flow." In: *Journal of the Atmospheric Sciences* 45.23, pp. 3588–3605.

Index

www.ingramcontent.com/pod-product-compliance
Ingram Content Group UK Ltd.
Pitfield, Milton Keynes, MK11 3LW, UK
UKHW052115210626
472488UK00005B/34
* 9 7 8 1 0 0 9 4 3 7 2 6 4 *